I. KHALIFMAN

A BOOK ON THE BIOLOGY
OF THE BEE-COLONY
AND THE ACHIEVEMENTS
OF BEE-SCIENCE

Laika Press
Chicago

Bees © Laika Press 2022
P.O. Box 3421
Chicago, IL 60601

The moral rights of the author has been asserted.

ISBN: 978-1-387-89155-9

Part of the *Soviet Science Series* published by Laika Press

www.laika.press

CONTENTS

Human reason has discovered many amazing things in nature and will discover still more, and will thereby increase its power over nature.

V. I. LENIN

PREFACE

Anyone making first acquaintance with bees finds himself in a world of amazing interest. This strange—one might say fantastic—world is magnificent in its organization and infinite in its variations, simple in its natural perfection and yet unbounded in its complexity.

Books on this wonderful world invariably shed light only on what distinguishes and differentiates bees from the rest of living nature. But if we examine bees from another standpoint, bringing out those traits which they share with the rest of the living world, if we investigate more closely those sides of bee life by which they resemble other animals and plants, the peculiarities of these truly unique creatures strike us still more forcibly.

Scientists are patiently pursuing their course towards mastering the secrets of Nature and subjugating her. Their discoveries flare the path of researchers into "bee-land." It becomes increasingly clear to anyone following scientists' progress, of what substantial importance are the phenomena of bee life they have discovered, and at the same time how much richer in contents than the wonders of bee life is the clear force of materialistic logic which embraces in planned experiment the minutest phenomena, forcing Nature to betray her innermost secrets for the service of man.

Today, bee-science is no longer what it used to be. The strict methods of biological analysis on which research has been based have armed experimenters with new facts. Precision technique in experiments and laboratory equipment has allowed deeper penetration into the substance of phenomena and processes. Thanks to this. new knowledge is being accumulated at a rate hitherto unknown.

The last discoveries of bee-science are a clear illustration of how the materialistic teachings of Pavlov, merging with Michurin's

7

teachings in the broad unity of Soviet agrobiology, are enriching theory and showing the way to the transformation of living nature for the benefit of socialist society.

* * *

Work on this book was started before the war at an experimental apiary in Zvenigorod, where research in connection with the training of bees was carried out under the direction of Professor A. F. Gubin. But not until 1949 was the book completed at the apiary of the experimental husbandry under the Lenin Academy of Agricultural Science in Gorki Leninskiye.

The illustrations in this edition are original drawings made by A. Sementsov-Ogiyevsky or drawings by other artists copied by him from various special publications dealing with the biology of bees.

The bees represented in the drawings are magnified as compared with the flowers.

FIRST ACQUAINTANCE WITH THE HIVE

NATURE AND MAN

A Peep into an Observatory Hive. What Was Found
in Spider Cave. How Bee Life Was Understood by Naturalists in the Past. What Enables the Soviet Man to
Observe the World with His Eyes Open.

An even and melodious buzzing is heard in a corner
of the laboratory throughout the summer. Here, in a narrow observatory hive placed sideways against the window
lives a small colony of bees. It is put here in spring, before the orchards are in bloom, and the bees soon grow
accustomed to their new, transparent home.

A broad glass lobby connecting the hive with the outer
world enables us to see its winged population hastening
all day long to and fro between their home and the outlet
in the window-frame.

Every moment bees take off one after another from
the wooden threshold nailed to the window-sill, soar buzzing into the air and disappear among the trees. A stream
of home-coming bees flies in the opposite direction. They
alight heavily on the threshold, hurry to the entrance,
then along the lobby to the hive, where they get lost among
the thousands of twin-like little creatures swarming on
the comb.

The nearer to the centre of the comb, the less we can
see of the geometrically-regular pattern of the cells covered as they are by a mass of scurrying and shuffling,
creeping and crawling bees. It is difficult to describe this
incessant and, at first glance, chaotic movement over the
immovable pattern of cells.

Some bees are crawling into empty cells, almost disappearing in them, others are walking as in a dream over the comb, others again are scrambling backwards out of the cells, on the bottoms of which lie the white circles of minute larvae, while yet others, agile and swift, are flitting post-haste by.

A fat drone lazily pushes his way among the inmates of the hive along the edge of the comb. The queen proceeds majestically, trailing her long abdomen, the bees making way for her.

Here a bee, returning home with coloured pellets of pollen stuck to her hind legs, is crawling up the comb; she runs from one cell to another in search of an empty one in which she deposits her burden with one swift

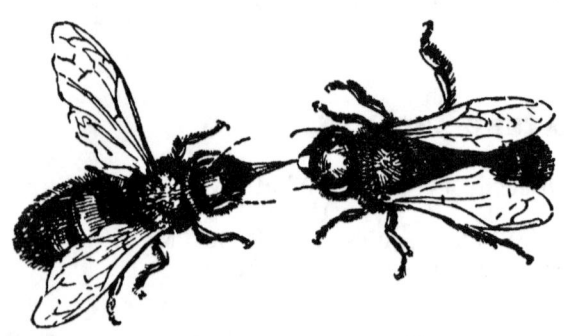

It is not very often that bees feed one another, still this can be observed at all hours of the day and in all seasons.... The ability to share food is an important trait of bee biology

movement. Another bee covered all over with pollen follows her and clinging with her legs to the sides of the cell begins ramming the pollen with her head, to press it into a tight mass in the cell.

There a scavenger-bee is staggering under the body of a dead wasp which she is dragging from the centre of the brood chamber. Passing through the entrance, she mounts into the air with her burden and flies some distance from the hive. Other scavenger-bees are sweeping the hive bottom with their wings, although it is already spotless. Not far from the entrance, a forager just back from the field is passing nectar to a home-bee. If we keep an eye on the forager we may see that she runs to the comb and in the very thick of the crowd starts circling about, folding and unfolding her wings.

Suddenly a high whining note cuts momentarily into the soft rustle of insects in the glass-walled observatory

—it is the fanning bees ventilating the hive. They stand with their stretched legs and their abdomens slightly tilted up, their four wings vibrating at such a speed that they are invisible.

Every corner of the hive is bubbling with life, but to one who has patiently watched the bustle on the comb, it no longer seems chaotic. One begins to understand that the thousands of four-winged insects in this community are connected by some inner ties.

The transparent walls of an observatory hive enable one to penetrate into many important details of the life of the bee community. But here, on a single comb, there are comparatively few bees: usually a hive contains a dozen, a score, or more combs. What, then, is the life led in their nests by these creatures that have interested man from time immemorial?

It is very, very long since man discovered bees. We think it worth while to tell our readers how this was established.

The upper part of the drawing on the wall of Spider Cave. For the bee nest the artist used a cavity in the cliff

North-west of the poor Spanish hamlet of Bicorp, in the mountains near Valencia, there are large deposits of brown and red hematite, yellow iron ochre, black manganese, and white lime. By pounding into a fine powder a piece of each ore and mixing it with animal fat, white, brown, red, and yellow paints are made. It is with such paints that the walls of Cuevas de la Araña (Spider Cave), discovered in 1919 near Bicorp, are decorated.

The wall paintings in that cave, as in many other caves of primitive man, represent human beings, animals, birds, and hunting. But Spider Cave offers us the first and the only yet known painting of a gatherer of wild honey.

Enormous aerial roots or giant stalks of some creeper plants hang from a steep cliff. A human figure, climbing up them, is scaling the face of the cliff. On the top, scarcely perceptible from below, is another human figure that has reached the entrance of a bee nest hidden in a niche, with enraged bees circling around.

We do not know what man in those times thought at the sight of a bee nest hidden among rocks or whether it reminded him of his tribe's dwelling. But we do know that people, not only of the slave and feudal epochs but also of the bourgeois epoch, saw in the life of bees the reflexion of their own social life.

And that was the case not only in the past. It is asserted in some modern works by foreign, for instance, American, bee-scientists that the well-regulated life in the beehive is the result of direction by a "secret bees' committee," a kind of unseen "board of the hive corporation." Allen Lathan, well known among U.S. apiarists, states quite seriously that the life of the beehive is governed by "control bees"—he probably means something like holders of control packets of shares; these bees are neither very young nor very old, but at what he terms "the golden age of the bee," and probably not very numerous.

Recently a certain F. D. Trollop-Belew published a book in which he asserted that life in the beehive is governed by three or four bees. These, naturally, take no part in physical work but devote themselves exclusively to organizing the honey and wax industries and co-ordinating the activities of the different groups of bees.

The latest editions of an American encyclopaedia of bee-keeping also instil into the readers' minds the idea that every hive has something like its own Wall Street, controlling "public opinion" among the bees, determining their tastes and their "home and foreign policy."

Bourgeois specialists of our day certainly have an incomparably better knowledge of bee biology than had any prehistoric honey-gatherer. Today scientists have at their disposal well-stocked libraries, special research institutions and laboratories, the data of related sciences, general and particular, perfected microscopes and most exact chemical analyses. But all this is not sufficient by itself to provide a clear insight into Nature, and a correct understanding of her.

However potent the supertelescopes through which the astral worlds of the universe are studied, however perfect the optical apparatus used for studying the microscopic world of the cell, however exact the methods of higher mathematical analysis enabling scientists to grasp phenomena and laws which cannot be observed directly—all this scientific equipment cannot by itself safeguard from errors of thought scientists brought up in capitalist civilization. Their psychology is still limited by class and social prejudices.

As Stalin said: "Whatever is man's manner of life, such is his manner of thought."

This law manifests itself, among other things, in the tendency of many bourgeois biologists to attribute to objects and phenomena of living and non-living nature qualities which in reality they do not possess and which, in the final analysis, are but a transfigured reflection of the relations of production prevailing in bourgeois society. Consequently, it was not just a coincidence that the Austrian zoologist F. Trögel innocently admitted in a book published shortly before the appearance of Darwin's *Origin of Species* that in the animal kingdom "the astonished observer sees everywhere an exact reflection of all our social, industrial, artistic, scientific and political life."

Even the greatest among the naturalists of the past were led astray by this illusory and distorted conception of nature.

Marx mentioned such a case in speaking about *The Origin of Species*, remarking that even a strict scientist like Darwin saw the world of man in the animal and vegetable kingdoms, in which he recognized his English society with its division of labour, competition, the discovery of new markets, different inventions and Malthus's struggle for life. And Marx repeated in his letter to Engels that Darwin's animal kingdom was represented as a civil society. Engels agreed with Marx and said jokingly that Darwin's teaching was a parody on bourgeois society of the time.

No wonder, then, that observers at different times found in the world of bees, with their various laws of life,

13

an exact reflection of the social order under which they were born and lived.

We know that the ancient Egyptians, for instance, regarded the bee-colony as a state headed by the pharaoh-bee who, surrounded by a retinue of servants fanning him with their feelers, watched from his lofty throne caravans of slave-bees laying their sweet tribute at his feet.

Later, Plato (IV century B.C.), and still later—Aristotle in his *History of Animals* saw in the bee-colony a slave society ruled by aristocrats—the drones.

The Romans corroborated the views of the Greeks. In his *Natural History* Pliny described a "spot of diamond-like brilliance" on the brow of the Caesar-bee, his splendour and proud carriage and the formidable guards around him. In Part Four of his *Georgics*, wholly devoted to bees, Virgil also asserted that in the bee-colony "the king of bees controls life."

Fifteen hundred years later, the bee-colony was still considered a monarchy, as may be seen from the following passage in *King Henry the Fifth*, where Shakespeare gives a colourful picture of bee life as understood in those days·

> *They have a king and officers of sorts,*
> *Where some, like magistrates, correct at home,*
> *Others, like merchants, venture trade abroad,*
> *Others, like soldiers, armed in their stings,*
> *Make boot upon the summer's velvet buds;*
> *Which pillage they with merry march bring home*
> *To the tent-royal of their emperor:*
> *Who, busied in his majesty, surveys*
> *The singing masons building roofs of gold,*
> *The civil citizens kneading up the honey,*
> *The poor mechanic porters crowding in*
> *Their heavy burdens at his narrow gate,*
> *The sad ey'd justice, with his surly hum,*
> *Delivering o'er to executers pale*
> *The lazy yawning drone. . . .*

But while in works of British authors of the XVI century the bee-colony looks ridiculously like the merchant Elizabethan England, French authors of the XVII century represented it as based on the classical feudal foundations.

Today one cannot help smiling on reading the work of the French writer Simon. He described a "bee state" in which the tired wanderers bringing goods from far-away are met at the entrance to the hive-city by guard-bees, and in which, before swarming, the bee-king notifies his subjects of the forthcoming sally by a flourish of a silver trumpet. . . . According to Simon, several kings can live in one hive, isolated from one another by the comb walls. He wrote further: "If one of the kings makes up his mind to extend his rule over the whole hive, jealousy is engendered among the kings, and faction and riot among the subjects."

The bee "monarchy" existed in bee-keepers' minds for a very long time and at each new stage in history was a more or less faithful copy of the corresponding human system.

The earliest work on bees in Russia was written by Pyotr Rychkov, an outstanding scientist of the Lomonosov school, who was an economist, a traveller, a geographer and an author. In his work, the bee-colony is described as something resembling the "enlightened despotism" of Catherine's time.

Late in the XIX century, a bee-keeper described the settling of a swarm in words taken from the coronation reports in *Politseiskiye Vedomosti*: "Calmly, with a dignity all her own, the queen enters the hive while the bees lined up on both sides utter a loud noise like the enthusiastic 'hurra!' with which the people cheer their tsar or tsarina."

But why speak of the XIX century? In our own time British bee-literature describes the bee-colony as a monarchy, and, of course, the peculiarly British brand in which "the queen has not even the rights of a constitutional monarch," but is merely "the flag on the battlement."

Yet by the end of the XVIII, and especially the beginning of the XIX century, the idea of bee life began to undergo certain changes corresponding to the changes in the way of life and of thinking peculiar to each individual country.

In his *Psychical Life of Animals* L. Büchner wrote:

"The queen is under the supervision and control of the workers. . . . She enjoys no inviolability of person, and both her throne and her life are dependent on the correct performance of her royal duties."

These first sallies against the "bee monarchy" and tales of the "popular rule of bees" were considered by some would-be researchers as dangerous sedition and an impermissible liberty on the part of trouble-makers.

P. Mikyalen-Mikalovsky, a Lithuanian landowner, indignantly exclaims in his book *The Bee*. "Bee-keeping, once so much in vogue, has become impossible since bees were discovered to possess constitutions, parliaments, and legal codes." And he was not the only one to shy at shadows of shadows and to demand that the works on bees by "Voltaire-minded" writers should be "locked up in the farthest corners of bookcases" so that "no servants should happen to come across them."

In unison with the Lithuanian provincial monarchist, Gaston Bonnier, a member of the French Academy, in a paper published in the *Revue Internationale de Sociologie*, bewailed the attempts at seeing in bee life "the realized ideal of collectivism."

It would be superfluous to quote here the numerous works in which the problem of "improving" bourgeois society was treated in the light of the "bee experience." Johannus Loccenius's book *The Community of Bees as Compared with the Civil State or: True and Thorough Example of Civic Life Derived from the Very Nature of Bees*, translated from the Latin and published in Moscow in the late XVIII century, cries shame on the filth of the human dwelling, on dirt in cities and villages and suggests following the example of bees in organizing communal sanitary facilities! M-me Clémence-Auguste Royer, the first translator of Darwin into French ("the ardent blue stocking" as she was referred to in the literature of the day), in a book published in Paris in the late nineteen-nineties, asserted that men were the cause of all human sorrow and drew a picture of an exclusively women's state organized on the model of the bee-community.

Along with amusing literary anecdotes like those quoted above, the works of Russian authors of repute made use of the example of bees to expose and castigate the abnormalities of a monarchist regime and a society duped and robbed by the "drones." Such were D. I. Pisarev's *Bees* and Leo Tolstoy's *The History of a Skep with a Bast Roof*. No wonder Pisarev's article was mauled by tsarist censors while Tolstoy's tale was published in full only in our time. But then Pisarev's article and Tolstoy's tale were didactic stories, political pamphlets. As for authors of works dealing solely with bee-keeping or bee-science, they all took great care to avoid analogies unacceptable to the ruling classes of bourgeois society.

Need we speak of the objective meaning and the purpose of the sociological interpretation of the biological phenomena that we find in the history of bee-science?

It is perfectly clear that the unscientific and warped interpretations of the life of the bee-colony by different authors and in different times have been and are attempts, often premeditated, to convince the readers that the existing social system of exploitation is in conformity with nature, is necessary and eternal.

All these ideas have often been expressed in a roundabout, disguised way and maintained by various means.

Maurice Maeterlinck, the author of *Bee Life*, which made such a stir at the beginning of the present century and was translated into many languages, stood in awe before a hive and, looking from bee to man, made a mental comparison between the world of men and that of bees; in confusion he recollected Robinson Crusoe as he saw the print of a man's foot in the sand. "Somebody has been here before," he thought.

Speaking thus of human society, Maeterlinck meant that in his evolution man had arrived at a stage long left behind by bees, that humanity had to travel a long and arduous road before it reached the level attained by bees in the organization of social life. Such are the ideas Maeterlinck propounds between the lines. In this allegorical disguise is presented the shabby idea so beloved of reac-

There are also a considerable number of books whose authors extol bees as "the biggest and most : vital philosophy in the world," and contend that "modern man, like man of the past, finds something hopeful in the solidarity of the beehive."

Let us see what kind of hopes are meant here.

One of those books insinuates that the time will come when wise men "will give the nations new laws patterned on the laws of bee life, and the golden age will set in on earth."

By trying to inspire the reader with admiration for the "wise natural laws of the bee state," the authors of such works want to make him believe that the natural course of evolution will finally bring humanity to something better, and that the social problem will be solved peacefully, without revolutions, and just as successfully as with the bees.

Thus both the "pessimists" and the "optimists" have one premise in common: people must not fight against the existing, bourgeois order of life.

An observatory hive is very useful for the bee-keeper as it enables him the better to know the state of his bees and comprehend the life of bee-colonies

Above: A simple model of observatory hive

Below: A latest model, in which the one-frame flat "observatory" part is connected with a super containing from six to eight frames

A seemingly peaceful branch of science—bee-science—turns out to be both an arena and a weapon of ideological struggle.

* * *

Let us try, with the help of the latest data of biological theory and practical bee-keeping, to find out here what the bee-colony actually is.

In the Soviet Union, where conditions giving rise to a distorted conception of nature and of society have been eliminated, man's eyes have been really opened to the world around him for the first time in history.

The Soviet man is vitally interested in seeing the world as it really is. By disclosing conditions that bring about various natural phenomena, we acquire a sure means of directing the comprehended natural process. Stalin stated that "man can discover laws, get to know them and master them, learn to apply them with full understanding, utilize them in the interests of society, and thus subjugate them, secure mastery over them." That is why the correct, materialist understanding of nature is a potent factor. The peoples of the U.S.S.R. use it to refashion the earth. This book will prove that bees, too, can and must be utilized for this purpose.

BEES ON DIFFERENT CONTINENTS

Bees of the European Continent. Red Indian Bees. The American Melipona. The History of Bees: an Example of Whole Species Eliminated in the Process of Artificial Selection. Why the Bee Scattered Throughout the World Remains the Least Variable of Domestic Animals. A Few Words about a Collective-Farm Apiary.

The discovery of the usefulness of the bee was an important event in human history. Only now can we correctly understand and assess the consequences of this important discovery, which was made at least four times.

The inhabitants of India and the other south Pacific countries have known the worth of indigenous Red Indian bees from time immemorial. Bees of this race are found living in a wild state in the taiga forests of the Soviet Far East. Indian bees, however, are not easily domesticated and that is the reason why only the honey-bee has been cultivated up till now.

The natives of America also had their own race of bees.

At the time the ships of European seafarers cast anchor off the shore of Central America, the Indians gathered honey from the Melipona, a tiny bee living in horizontal combs resting on wax columns. The Melipona is dark-coloured with white bands and is covered with golden fuzz; it uses combs for rearing the brood and stores honey in special "jars" made of dark wax possessing curative properties.

The Melipona belongs to a genus of the Apidea which is rich in species. Here is a worker Melipona

Had there been no bees in Europe—and European bees are better in many respects than the Melipona—the conquerors of America, together with gold and precious stones taken from the natives, would no doubt have brought to Europe, as a curiosity, the pumpkins in which the Red Indians kept the Melipona. Then, perhaps, she would be flying all over the world as is now the honey-bee.

But it was exactly the opposite that happened.

A century after America had been discovered, the first colonies of honey-bees were taken there; these quickly acclimatized on the new continent to the almost complete exclusion of the Melipona from bee-keeping.

According to Darwin, the indigenous bee races in Australia, too, were superseded in bee-keeping by the European bee.

Thus we see that unconsciously and without clearly perceiving the importance of the bee, different peoples living on different continents, at different times and quite independently of one another, chose and utilized this insect.

But there is a curious fact about bees, which naturalists noted long ago: whereas all plants and animals for
any length of time domesticated and tamed by man have
radically changed under his influence, bees seem to remain
unchanged.

Today, varieties and breeds
of plants and animals reared
by man are incomparably more
numerous, and the differences
between them much more
strongly marked than the differences in the forms which in
nature are regarded as different species and even different
genera. It would be enough to
call to mind horses, sheep,
dogs, poultry, pigeons, and
canaries, with the striking
characteristics of each breed, so
forcibly revealing the almost
limitless variability of domestic
animals.

A bronze hive from Pompeii,
a relic of Roman apiculture

We should bear in mind that it is not long since the
breeding of canaries was taken up. Even pigeons became
of interest to man at a much later date than that at which
the unknown artist painted the honey-gatherers on the
walls of Spider Cave.

Why, then, do only the natural, geographic races of
the honey-bee exist today, and why is there not a single
"artificial" race bred by man on the principle of selection?

Why is the honey-bee practically the one and strongly
marked exception among the mass of plant and animal
species radically changed by man's effort?

Many biologists assert with good reason that the bee
is in a *semi-domesticated* state and that centuries of bee-
keeping have had little effect on her.

A swarm absconded from an apiary is perfectly at
home among wild nature and in no way suffers from the
absence of human supervision. At the same time, if we

find a stock in the hollow of a century-old tree in a dense forest and carry it to an apiary, the most observant of bee-keepers will not be able to detect any essential differences from the hive bees either in the bees' anatomy or the main lines of the inner life of the colony.

Above: A Mexican earthen hive for the American Melipona

Below: A bark hive for the African Melipona, from Tanganyika

These differences seem to manifest themselves in one respect only: in the attitude of wild bees towards man. The mountainous-taiga bees recently discovered in Bisert and Shalya districts of Sverdlovsk Region are reported to be extremely vicious. Even if one succeeds in driving a swarm flown from the hollow of an old tree into a skep usually hung on a tree-trunk, one cannot be sure of being able to take the skep from the tree; only a very skilful bee-hunter can take possession of the swarm. More often than not the unlucky bee enthusiast is driven off the tree to the amusement of experienced bee-keepers.

With these bees you cannot be sure even of the swarms you have hived. It is no use putting them into a modern frame hive straightaway: it is thought advisable to keep the bees in skeps for a year or two. These bees are very impatient of the bee-keeper's attempts at interference in the life of the colony and often react to such attempts by flying home—to the taiga.

In other aspects, wild bees are not very different from domesticated bees which have long existed under the bee-keeper's care and for many generations inhabited frame hives, building combs on factory-made foundations. So bees remain unchanged while almost all cultivated plants and animals have been so greatly changed through artificial selection that it is hard, sometimes even impossible, to recognize their wild ancestors.

Charles Darwin acknowledged the bee to be "the least variable of all domesticated animals" and explained this by the fact that while living in the apiary the bee "feeds itself and follows in most respects its natural habits of life."

It would be highly instructive to understand what these natural habits of bee life are, to study them closely and discover such of their peculiarities as, having made the bee the least variable of all domestic animals, have enabled man from decade to decade to spread bees, in their present semi-domesticated state, throughout the five continents, throughout all countries and zones from the subarctic to the tropics.

The U.S.S.R. has long and firmly held the first place in the world as regards the number of apiaries.

In the central forest and forest-steppe zones of the Soviet Union, black forest bees are kept, in the southwest, the Ukrainian steppe bees, in the south, the famous ancient Caucasian bees—the mountain and plain varieties, grey and yellow.

All these are "geographical races," the natural breeds of the honey-bee.

They differ both in size and colour and in certain noticeable anatomic characteristics and behaviour. For instance, the Caucasians are gentler than the forest bees. And then we have a good reason to think that the relatively greater docility of the Caucasians, whose domestication was supposedly begun at a much earlier date, is a characteristic bred through training and artificial selection.

As for the natural differences between breeds they are the result of different natural conditions and are of an adaptive nature. In northern regions, for instance, where the honey-flow period is shorter and flowers secrete more nectar, the bee's proboscis is shorter. In those regions the annual fluctuations of temperature are more abrupt

The contrivances in which man keeps bees are infinitely varied.
1) Old Russian hollow logs. West European wooden log-hives covered
with boards (2) and with stone flags (3). 4)Old French skeps. 5) Willow-
twig skeps. 6) Lath skeps. 7) Straw skeps. 8) Alpine earthen crocks,
with entrances. 9) Mud tubs in which bees are kept in Egypt. 10 and
11) Abyssinian sewn hives

and the capping of ripe honey in the combs is done more elaborately than in the south: in capping honey, the northern bees leave a space so that the changed volume of the honey under the influence of temperature does not break the cap. . . .

Before proceeding further in our narrative, however, let us pay a visit to an apiary.

Dozens of standard hives painted white, blue, and yellow are dotted over a lawn surrounded by willows, maples, and lime-trees and lying well away from the dwellings and collective-farm service buildings. Control hives stand on weighing machines in sheds. On a low platform in the centre is a keg from which water drips along a groove in a sloping board for bees to take water.

The air warmed by the scorching rays of a summer sun is filled with the heady odour of honey,

House-apiaries are widespread in the world. Southern houses, like the Turk house above, are in every respect inferior to northern, like the German house-apiary below

flowers and wax, and rings with the incessant buzzing of flying bees.

A man in a broad-brimmed hat from which a thick black veil hangs on to his shoulders checks the number of a hive and puffing a whiff of smoke into the bee-entrance, removes the outer cover. He takes away first the cushion and then the inner canvas cover and puffs a bigger volume of smoke, which makes the forty or fifty thousand bees inhabiting the hive rush to the combs and cling to the honey-filled cells. This seems to be an age-old instinct: sensing the smoke of a forest fire, bees living in hollow trees filled their crops with honey before fleeing.

While the bees are sucking honey the bee-keeper swiftly but unhurriedly removes the light wooden frames with the combs from the hive body and examines them one after another. The cells are an open book for a bee-keeper, and almost imperceptible signs and indications tell him of the state of the colony and of its needs.

The knowledge of the laws governing bee life prompts him to take such measures as will make the multitude of diligent and self-willed insects in each colony do his will, not merely applying themselves more vigorously to building wax combs and filling them with honey, but flying to forage where the agronomist wants them to.

UNDER THE MAGNIFYING GLASS

THE HONEY-BEE AND HER TOOLS

A Survey of the Anatomy of the Bee. The Difference
Between a Young Bee and an Old One. Two Types of
Bee's Eyes. The Mouth, the Proboscis and the Tongue
with the Spoon. The Antennae. The Flight of the Bee
and the Peculiarities of the Body and Wings. The Organs of the Bee Are Her Instruments. The Honey-Sack
and the Stomach-Mouth.

The bees we usually see are what we call worker-bees,
or rather, grown-up workers. In steppe regions they fly
four or five kilometres, and sometimes even farther, away
from home and may be seen in the most secluded spots
where a nectareous plant is in bloom.

It is but very seldom that a stranger has a chance of
seeing the male bee—the drone—outside an apiary. Still
more seldom has one the honour of beholding the female
bee—the queen. Although she is the longest-lived of all
the bee-stock, the queen leaves the hive not more than
three times a season, and that only for a few moments.
Her first outing is an orientation flight, to get to know
the surroundings, her second, the mating flight, and the
third is with a swarm, when part of the colony leaves its
old home.

As for the young bees, only the bee-keeper sees them.

Outwardly, however, a young newly-born worker differs little from an old bee. A bee is never young or growing.
She is born a perfect full-sized insect and the only evidence of her youth is that she does not leave the hive, so
when a bee takes her first flight it is a sign of her maturity.

The drone and the queen differ markedly from the
worker, and we shall study them apart. Now we shall speak
of the worker.

A worker is from twelve to fourteen millimetres long and five to six millimetres high. She weighs about 0.1 gramme when empty and up to 0.15 gramme full. Often bees have to carry great weights: a bee flying from the hive with a dead drone carries almost 0.2 gramme, twice as much as her own weight.

A chemical analysis of the dried body of a bee shows that it contains about 20 elements, including aluminium and fluorine, but mostly carbon, nitrogen, and oxygen in complex combinations with sulphur and phosphorus.

Modern movable-frame hive has greatly lightened manipulation and enabled man to control the growth and development of the bee-colony. Depending on the number of frames in the hive body, hives are divided into vertical and horizontal. Here is a horizontal hive

The small and fragile bee in her resilient thin chitinous test deserves attention both as a flying apparatus and a chemical laboratory. The more or less watery nectar she sucks from the flowers undergoes changes in the flying bee, becoming half-finished honey which will be ripened in the hive.

Pollen gathered from the stamens of plants is mixed on the leg of the bee with a small quantity of honey and is transformed into a chemically different paste which in the hive will become "bee bread," a very important item of the diet of both adult bees and brood.

A dark head with two long antennae or feelers consisting of twelve joints and set on the thorax by a white slender and flexible neck (seen only at close quarters when the bee lowers her head), a darkish thorax with two pairs of transparent wings and three dissimilar pairs of legs, an ever-moving abdomen—this seems to be all there is to a bee at first glance. Such a casual glance does not reveal anything remarkable in the insect.

But it is worth one's while to study a bee under the magnifying glass.

The triangular head of a bee is covered with greyish fuzz. Two big black eyes, each consisting of about 5,000 tubular units—facets—protrude like light-focusing lenses on each side of the head. Research has shown that the optical impressions received by these eyes are made up of separate points, like photographs reproduced in books. They belong to the category which specialists call compound eyes. In the top of the head are located three small simple eyes, the ocelli.

A bee leaves her dark hive on a bright sunny day and flies looking with the ten thousand unblinking side facets and the three "cyclop" eyes in front. By observing the conduct of bees whose eyes had been painted over with light-proof varnishes, it was established that the bee sees with the help of the compound eyes, the ocelli merely contributing to her general sensitiveness to light.

It is interesting to note that the eyes of young bees see differently from those of old ones: the older a bee, the more sensitive she is to light and the more strongly she is attracted towards it.

Bees can recognize colour, though this ability is not well developed.

If, for instance, ten sheets of paper of different colours—black, white, red, pink, orange, yellow, green, blue, violet, and light-blue—are put on a table in the vicinity of a hive, a bowl with syrup being placed on one of them and similar bowls of water on the others, the bees will soon fly for the syrup and will unerringly find the right bowl.

Suppose the syrup is left for some time on the blue sheet, so that the bees may remember the colour associated with food. If the syrup is then transferred, say, to the yellow sheet and a bowl of water is put on the blue, the bees will continue to fly to the blue sheet, even if the order in which the differently-coloured sheets are disposed on the table is changed in different ways.

On detailed study of the colour-sensitiveness of bees by this method, it was discovered that they are insensible to red and confuse it with dark grey, while green is confused with blue and yellow. A long series of new and clever experiments (particularly experiments with artificial cloth and paper flowers) helped to establish that bees can see white, yellow and blue, but distinguish other colours only by their intensity.

But bees' eyes see something that is hidden from the human eye. We can study the world by the light of the ultraviolet part of the solar spectrum only with the help of photographic plates covered with a special emulsion, while bees perceive it with their eyes.

Experiments were made to study how bees distinguish simple colours and combinations of contrasting colours (blue against grey, grey against yellow, yellow against violet) which definitely established that bees see red flowers as grey, purple as blue, white as green, and green as yellowish. Thus a bed of red poppies bordered with white daisies seems to bees almost black in a border of green, while the surrounding grass seems light-yellow.

The eyes are located in the upper part of the head; attached to the lower part is the mouth apparatus consisting of four jaws—the two mandibles and the two maxillae, the latter forming part of the labium. In spite of the number of jaws and contrary to the prevalent opinion, a bee is practically incapable of biting through the skin of fruit. The jaws work sideways, like double pincers. The long proboscis projects downwards, and is straightened out in action by a movement like the unfolding of a penknife. The bee sucks nectar from the bottom of deep flowers with her proboscis as with a straw.

If the nectar is thick the diameter of the canal formed by the complex mouth is increased. When the flower contains little nectar the bee laps it up with the "spoon," or labellum, at the tip of the tongue. The tongue itself is hairy and curls up like a thin worm, it is almost half the bee's body in length and, what is most astonishing, is quite red.

The bee can take in dry food, too. She wets sugar with saliva or water and then sucks the syrup with her proboscis.

On the whole, the bee's mouth is highly perfected. The complex movements of the jaws, the proboscis, and the tongue enable the bee to lap, to lick and to suck food.

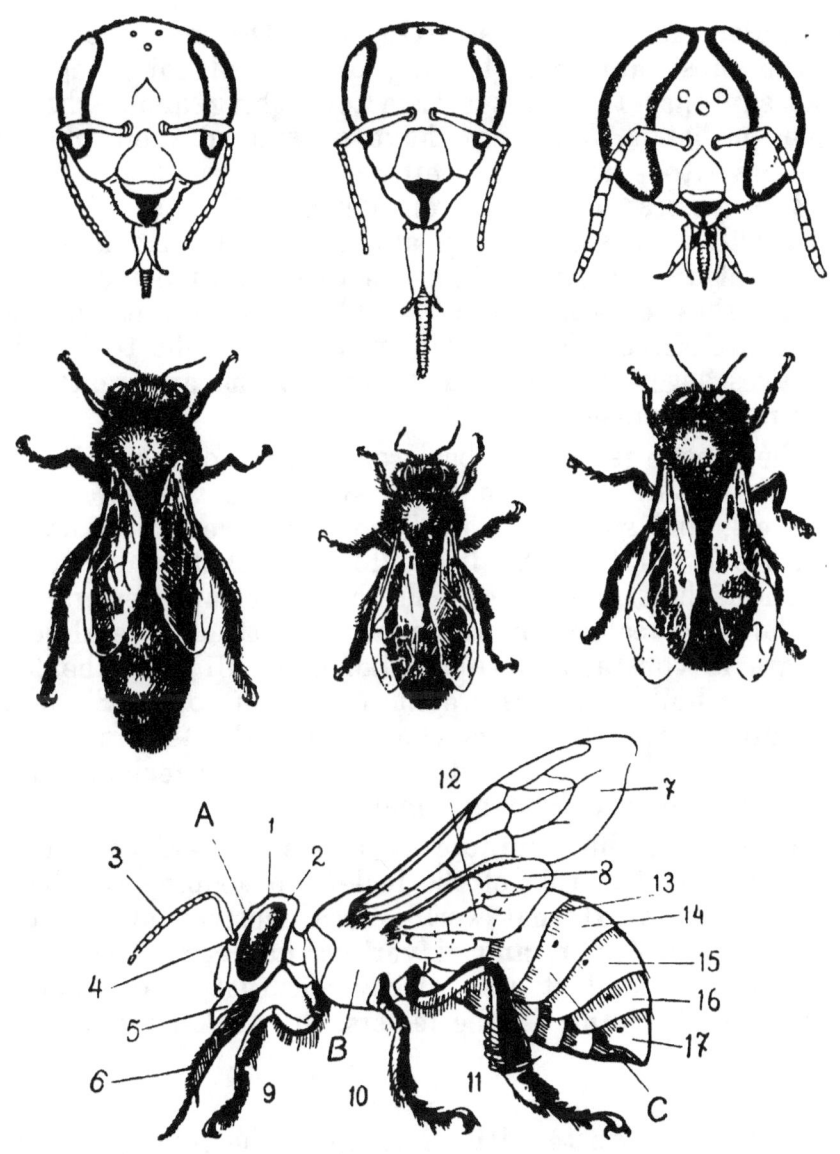

Centre, left to right: Queen, worker-bee, and drone. *Upper row*: It is obvious that the heads of the three castes of the honey-bee differ not only in shape but in the size of the compound eyes, in the situation of the ocelli, in the shape and number of joints in the feelers, in the length and construction of the proboscis, in the shape of the jaws, etc.

A. Head: 1. Ocelli. 2. Compound eye. 3. Antennae or feelers. 4. Labrum 5. Maxilla. 6. Proboscis.

B. Thorax: 7. Fore wing. 8. Hind wing. 9. Fore leg. 10. Middle leg. 11. Hind leg.

C. Abdomen: 12-17. Segments of abdomen.

Anatomists are right in asserting that the mouth of a bee is "the most universal feeding organ," although agronomists say that the least satisfactory thing about the bee is her proboscis. When we have considered their plea we shall have to agree with their views.

Bees possess a fairly well developed sense of taste. A bee will suck sugar syrup and quite correctly assess its concentration, preferring thicker solutions. Like man, a bee distinguishes between acidity and saltiness but will not touch saccharine, sweet as it may be thought to be. Yet she takes sugar mixed with bitter quinine apparently not minding the bitter taste.

The tongue is not the bee's only organ of taste: if a bee puts her leg in a drop of sugar syrup she will stretch her proboscis, and suck it, which she will never do if her leg is in a salt solution. So it would seem that under certain conditions bees can taste with their legs and feelers.

The long feelers, or antennae, consist of several joints and are in constant motion. Like hairs all over the bee's body, the hollow feelers are the organs of touch. Six thousand plate organs covering them make the feelers the organs of smell as well, and bees with their feelers cut off are unable to find food by smell.

A series of most minute experiments were carried out to investigate the role of the feelers in keeping the bee in touch with the external world. Bees deprived of both their feelers, of one feeler, and of various numbers of the joints were studied, and it was discovered that the fewer the number of the joints in the feelers, the weaker the olfactory sense of bees.

So bees can taste with the help of the tongue, the legs and the feelers.

As far as sound is concerned, it has not been discovered what organs are responsible for conveying auditory impulses to bees.

Some bee-keepers think, therefore, that bees are deaf. Others, on the contrary, are quite sure that bees can not only hear, but also perceive minute vibrations of objects and of the air.

Indeed, bees have no known organ of hearing. But if the sound made by the young queen just emerging from her cell is reproduced by the vibrator the bees on the combs immediately react to it: they become still just as at the "voice" of a real queen. Moreover, mature queens in their cells respond to this sound. So it is clear that the sound signal produced by the vibrator is perceived by the bees.

But this is not all either.

Recently an apparatus registering ultra-short sound vibrations made it possible to hear the voices of various bee-colonies which are beyond man's power of perception. The soundless singing of the mass of bees come to the feeder with honey, the signals of the bees jostling one another at the entrance and on the alighting-board, and even the route signal sent by a flying bee could thus be heard.

Having started studying the ultra-high-frequency sounds made by bees, investigators have entered an unknown world, in which, no doubt, quite unexpected discoveries are in store for them.

The closer we study the working of the sense organs of bees, the more clearly we see all the marvels of the world in which these four-winged creatures live. In this world everything is unexpectedly displaced, everything bears the imprint of fantastic originality. Here, the taste of food can be distinguished by the legs and some researchers prove that the olfactory and tactile irritations received by neighbouring nervous cells are blended into one sensation unknown to man. Voices and noises perceived by our hearing are soundless to the bee which, on the other hand, hears ultra-high-frequency sounds which are beyond the range of human perception. Even the usual colours of sky and earth seem different to bees from what they seem to man. The sun itself seems different to them.

And yet man successfully investigates this non-human world and discovers things he did not know and did not suspect, and which he neither hears nor sees nor feels.

Engels wrote that ants have eyes different from ours, and that "the very fact that we are able to demonstrate that ants can see things invisible to us, and that this proof is based solely on perceptions made with *our* eyes, shows that the special construction of the human eye sets no absolute barrier to human cognition."

Neither does the specific construction of the human ear, as we already know, set an absolute barrier to human knowledge.

Marxist philosophy proved long ago that despite the relative imperfection of man's sense organs, their testimony is perfectly reliable, and that, basing himself on this testimony, man can confidently proceed along the road of knowledge. Dialectical materialism tirelessly exposes all and sundry inventions of idealists advocating the impossibility of knowing the objective world.

To study things, to penetrate into the essence of phenomena, to disclose the nature of processes and the moving forces behind them, requires the act of thinking, which, based on the evidence of the senses and controlled by practical experience, is a far-seeing, sensitive and acute instrument of cognition.

In the process of cognition, the ability to reproduce natural phenomenon, to bring it into being out of its conditions, has been and is still a most potent and convincing method of establishing the correctness of man's knowledge, of verifying the truthfulness of his conception of the essence of objects and phenomena.

This is why we have every reason to regard the knowledge obtained through research and throwing light on the laws governing the seemingly fantastic world of bees as perfectly reliable.

But nature never reveals her secrets herself.

To disclose the laws of living nature, a great sacrifice of labour, patience, and persistence is often needed, and thousands of correctly conducted experiments as well as thousands of successfully accomplished investigations offer us an example of lofty exploits of intellect, real triumphs of will, born of a noble striving after exact knowledge. The history of science knows a great number of experiments amazing by their subtlety of conception, high technique of execution, in themselves a perfect product of thought, beautiful and inspiring like true works of genius. Let us just recall Kliment Timiryazev's experiments in which he studied the role of different parts of the spectrum in the assimilation processes occurring in the chlorophyll of a green plant. Let us recall the experiments of Pyotr Lebedev, the physicist, who weighed the pressure exerted by the sun-ray, or Ivan Pavlov's, who registered the minutest changes in the nervous system of an animal he was experimenting upon by the number of drops of the gastric juice flowing from a pipe inserted into the animal's internal organs.

These exploits of intellect, these triumphs of will are necessary not only for research the results of which shed light upon vast regions of facts and phenomena. They are necessary for solving minor, particular, problems, the solution of which is the day-to-day work of science. Research workers had to carry out not a few experiments of the latter kind in studying the flying qualities of bees, in analyzing the structure of their flying apparatus.

What enables a bee to fly?

Her thorax is much smaller than a pea and its upper and nether parts are covered with chitinous segments within which is located a system of muscles which set in motion six legs and four wings.

The bee's four wings can carry her in the air at a rate of up to sixty-five kilometres per hour, i.e., over a kilometre a minute. With a full load the bee flies more slowly, covering a kilometre in about three minutes.

The transparent and pearly membranous wings of bees are strengthened by tubular veins running length- and breadthwise, which constitute the framework of the wing. When at rest, the wings are folded in two layers horizontally from thorax to abdomen, parallel to the longitudinal axis of the body.

When in flight a bee's wing makes over 25,000 flaps per minute, that is, 440 per second. The wings themselves are muscleless and are hinged by their bases to the mesothorax and the metathorax; they are moved up and down by the ends of thoracic segments, acting as powerful levers.

The segments are set in motion by two sets of muscles: one set running lengthwise along the thorax, the other running obliquely, both sets having very few nerves. Usually, the more active the organ, the denser the network of nerves. Research workers were puzzled how the wings could work so intensively when there were so few nerves in the muscles. For nobody had observed in living nature neuro-muscular structures capable of working at a velocity approximating that of the contracting muscles of an insect.

Scrupulous anatomical research helped to solve the riddle: it was seen that the contraction of the vertical thoracic muscles causes the conjugate oblique muscles to extend, and vice versa. The function of the nerves here is only to keep the muscles active. To break up the work of the wings into separate movements, a film camera with a capacity of more than a thousand exposures per second had to be used.

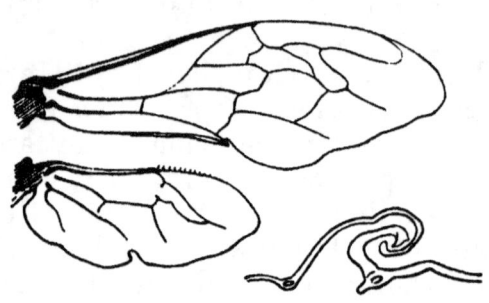

Right fore and hind wings of the bee. The upper end of the hind wing carries a series of minute hooks by means of which the fore and hind wings are coupled in flight, making a single aerial sail. Fanning bees work with their wings uncoupled. *Lower right corner*: The hooks

The analysis of the film showed that, preparing to fly, a bee extends her wings forward, at the same time drawing the fore wings over the hind wings and coupling them into two triangular sails at almost right angles to the thorax.

The study of the wing movements explains to us why a bee can immediately take off into the air like a helicopter. For bees foraging on flowers the ability to do so is very important.

A bee holds her body differently in the air during flight, depending on different conditions. She uses her 9-jointed legs to shift the centre of gravity: when empty, the hind legs are stretched back, when loaded, they are drawn forward to counterbalance the displacement of the centre of gravity.

On her way back to the hive the forager brushes pollen from her body and legs (using her left leg to brush her right side and vice versa), presses the pollen and deposits it in the so-called pollen baskets, hairy spaces on the hind legs.

The pellets of pollen on the bee's legs look exactly like little baskets and this is why the bee was often pictured in fairy-tales as a housewife hurrying home from the market with two baskets full of purchases.

A bee makes a six-point landing. When she walks (she does not crawl) she rests her weight on three points, at each step moving two legs on one side and one on the other. A walking bee is capable of developing considerable force: over a rough surface she can pull a load twenty times her own weight (a horse, it should be noted, usually pulls a load equal to its own weight).

The claws on the last segments of the legs enable a bee to walk easily on vertical surfaces such as the combs or the stalk of a plant, while the sticky pads between the claws adhere to smooth surfaces and enable her to move quickly over a vertical sheet of polished glass or the ceiling—upside down.

The structure of each pair of legs, the shape of each segment, the position and number of spines and hairs on them, show that these organs are tools common to plants and animals and created by nature in the process of evolution.

Kliment Timiryazev wrote: "The word organism includes the notion of tool, and implies the notion of use."

It would be interesting to study the legs of the bee from this angle.

The bee's legs are of use to her not only for locomotion, but for

Left foreleg of worker-bee. *Above:* face view and *below:* dorsal view. In the centre are (enlarged) the clasp-like lobe at the end of the tibia and the notch in the basal part of the basitarsus constituting the antenna-cleaner. In figure on page 160 the antenna-cleaner is shown in action

collecting food, building combs, and cleaning her body. These processes clearly show that the legs are work tools. Engels called bees "productive animals with tools." Observation proves that the shape of various spikes, brushes, combs, eyelets, spurs, tweezers, the nature of the joints and the different curvatures, go to make the legs highly developed work tools of almost universal application.

It has been established that the brushes of stiff hairs on the fore legs are used for removing pollen from the head and cleaning the compound eyes. When used for this latter purpose the brushes fulfil the function of eyelids.

Antenna-cleaners, consisting of a deep semi-circular notch on the basal part of the fifth segment (the basitarsus) and of a small clasp-like lobe that projects over the notch from the end of the fourth segment (the tibia) are situated in the inner margins of the fore legs, just beyond the tibiotarsal joints. The margin of the notch is fringed with a comb-like row of spines through which the antennae are passed. The long spines at the end of the fourth segment (the tibia) of the middle legs are used for loosening the pellets of pollen from the pollen baskets of the hind legs into the combs. Pollen presses for the transfer of pollen from the collecting brushes to the baskets are situated between the fourth (tibia) and the fifth (tarsus) segments of the hind legs.

The hind tibia is fringed with long, curved hairs and the space

Left middle leg of worker. *Above*: face view and *below*: dorsal view. The bee uses the long spine at the end of the tibia to loosen the pellets of pollen from the pollen-baskets into the cell

enclosed by these is the pollen basket, or corbicula, which we have already had occasion to mention. But this is not all: on the hind legs one can easily see a row of spikes, called "wax-shears," with which the young bee quickly removes pentagonal wax scales secreted by the glandular cells in the lower part of the abdomen.

The chief organ of circulation—the tubular heart, for which there is no room in the thorax—is situated in the abdomen. The blood, or rather haemolymph, makes up about 30 per cent of the entire weight of the bee. It contains about three per cent sugar, and in some periods of the insect's life, even more.

Bees also breathe with the abdomen through several pairs of breathing pores —the spiracles—which let the air into air-sacks situated in the head, the thorax and the abdomen, and into the tracheal tubes distributed throughout the body. These peculiarities of the bee's respiratory system enable her to suck up nectar without stopping to take breath.

Constantly expanding and contracting the abdo-

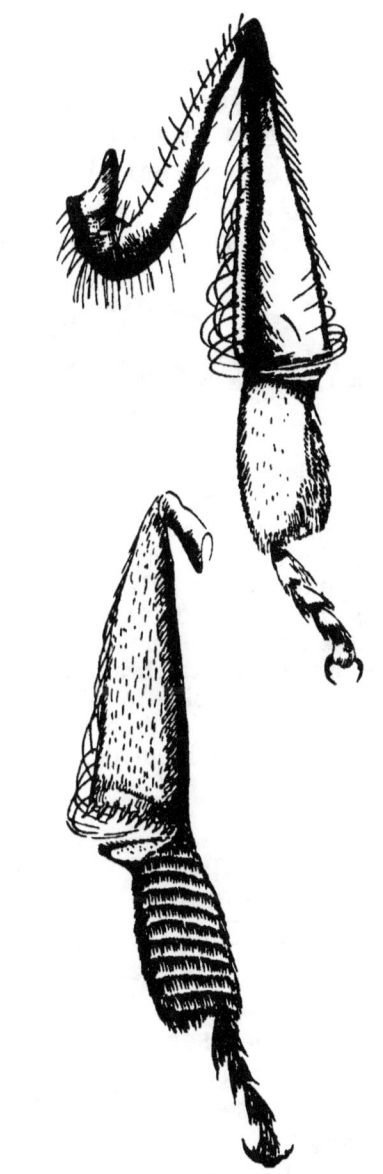

Left hind leg of worker. *Above*: fore view and *below*: dorsal view. One can well see the long curved hairs and the short stiff spines, as well as the "pollen press" between the tibia and the basitarsus. The figures on page 182 show how, with the help of the pollen press, pollen grains are scraped off from the body and formed into a pellet

men—one hundred and fifty movements per minute—the bee pumps the air into the air-sacks. The quantity of air needed by bees differs according to the season, being in winter fifteen times less than in summer.

The honey-sack is situated in the abdomen, beside the heart. Whenever the least opportunity presents itself, a bee will greedily suck up food and fill her honey-sack

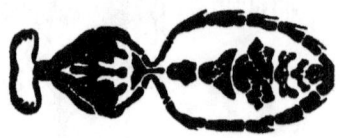 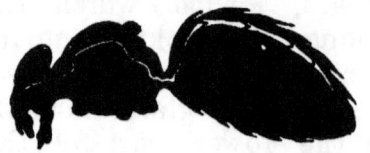

In the diagram of the respiratory system of the honey-bee as seen from above we can distinguish the thoracic and abdominal spiracles and also the air-sacks in the head, the thorax and the abdomen. The innumerable branching tracheae going from the air-sacks to all the parts of the body and entwining them are not designated in the diagram

The main organ of blood-circulation, the tubular heart, extending through the whole body of the bee is shown here from the side. The white jointed line in the abdomen shows the slits in the ventral part of the dorsal vessel – the ostia. The anterior part of the vessel ending in the head is the aorta

to its full capacity of up to 80 cubic millimetres of honey. With her abdomen distended, the bee is a real aerial honey tanker. But facts show that it is by no means gluttony that makes the bee suck up the food with such avidity. This can be seen from the mere fact that the inner wall of the honey-sack is lined with nectar-proof chitin.

This anatomic peculiarity must be borne in mind, for, as will be seen later, it is important for understanding bee biology.

At the very end of the abdomen is hidden the sting so familiar to many. But not everybody is familiar with the fact that this weapon of defence, which in past ages used to be the bee's ovipositor, carries ten recurved barbs.

These barbs, which prevent the bee from removing the sting from her enemy's body, once more bring into evidence the fact that every detail of the organs of a living body is a work tool. Owing to perfected methods of study of both tissues and individual cells, a vast number of such subservient devices visible only with the help of the microscope, have been discovered. They have been found in the cells of the trachea and tracheole, in the tissues lining the alimentary canal, in the peculiarities of glandular cells and in the cells of various parts of the integument. Everywhere the scientist's penetrating gaze discovers that the structure of any particle of a living organism necessarily conforms with its function. Each new discovery adds its weight to Engels' statement:

"The whole of organic nature is one continuous proof of the identity or inseparability of form and content. Morphological and physiological phenomena, form and function, mutually determine one another."

The bee-colony is always wide awake. Thousands of observers have studied it attentively and none has ever seen it asleep.

Perhaps this was because the bee-keeper always inspects the hive by light. Perhaps we waken bees by looking into the hive in day-time with the sun shining or at night with a lamp, thought the bee-students at an apiary near Moscow. What if we observe bees at night and in complete darkness?

Accordingly a few workers of a colony living in an observatory hive were marked with luminous paints. At night, one could see light dots moving on the dark combs, disappearing and reappearing among a mass of invisible dark bees, which proved the correctness of the old opinion that whether by day or by night, in bright or rainy weather, on a warm or a cold day, the bee-colony is always in motion.

Food to be looked for or collected, the brood to be fed, water to be brought to the hive or got by condensation from the nectar, combs to be built, the brood nest and the cells to be cleaned, the hive to be ventilated, the temperature in the hive to be lowered or raised—there is always something to do in the bee-colony. Some of the processes are carried out only by day, others, mostly by night, others again all around the clock; some of them follow one another, others are done simultaneously. But everything that is done in the hive is done by the workers with the single exception of egg-laying, which is the function of the queen alone. And once she has deposited an egg in the cell, the queen bothers no more about the progeny.

The best known function of worker-bees is prospecting for sources of food and collecting food. But, as will be seen later, the performance of this duty occupies comparatively little time in the bee's life

A worker does not lay eggs, her reproductive organs being so underdeveloped that we can consider her practically sexless. At the same time, not all sex properties are extinct in her: it is the worker-bee that rears the new generation. The secretion of their broodfood, or pharyngeal, glands (some hold that they are maxillary glands, others that they are glottal glands) is special food with which the larvae are fed. It would be wrong, therefore, to regard the queen as the only mother of bee generations. The saying "the mother is not she that gives birth, but she who rears" is especially true in this case.

If a colony is deprived of its queen through some accident, the young bees will soon have no brood to feed, and this affects them in such a way that they acquire the ability to lay eggs themselves. But such bees lay eggs in a haphazard way, sometimes several in one cell, and from their eggs develop only male bees—drones. The fact that an unfertile worker can become "semi-fertile" affords fresh proof that changed metabolism changes the properties of a living organism. We should add, however, that the worker-bee's ability to lay eggs is not the most astonishing example of such influence afforded by bee biology.

THE QUEEN AND HER RETINUE

The Properties and Behaviour of the Queen Distinguishing Her from Worker-Bees. The Queen's Functions and the Development of the Colony. More Observations on the Formative Qualities of Food. How the Brood Is Protected against External Influences. What Takes Place in a Queenless Hive.

Should a very experienced bee-keeper be asked to point out the queen in the brood-nest, he would not be able to find her at once among the mass of workers on the combs.

There are tens of thousands of worker-bees in a colony and only one queen. But she is easily recognized at first glance, for the queen is one and a half times or even twice as big as the worker. The queen differs from a worker-bee by her more round head with bigger compound eyes set wider apart; the ocelli are situated more in front, the twelve-joint antennae, too, are somewhat different; she has no wax glands and her legs have no pollen-carrying devices; her ovipositor is a curved four-barbed sting which the timorous queen seldom uses against external enemies. Her long tapering abdomen, containing two ovaries of from one to two hundred egg tubules or ovarioles each, protrudes far under the folded and comparatively weak wings.

Fifty or sixty hours after emerging from their spacious acorn-like cells on the lower edge of the comb or among worker cells, young queens have been known to make their pre-nuptial orientation flights, during which

Queen-bee (centre) surrounded by her retinue of young nurse-bees always accompanying the queen in her rounds of the combs

they familiarize themselves with the situation and surroundings of the hive. A few days later the virgin queen goes out on her mating flight and soon after impregnation begins to lay.

In a normal colony the queen lays eggs every day from the end of winter to the end of summer. At first the number of eggs runs into dozens a day, then hundreds, and at the height of the brood-rearing season the queen's daily output—1,500 eggs—is equal to her own weight. From this we can judge of the intensity of the metabolic processes occurring in her body. Within an egg-laying season a queen may lay from 150,000 to 200,000 eggs with a total weight a hundred times her own.

All day and all night the queen goes over the combs looking for good empty cells where she thrusts her head now and again, as if the better to ascertain their condition. Then she stops, curves her body forward and inserts the lower part of it into the cell, turning halfround upon herself clock-wise as if screwing her body into the comb.

This is a reproduction of the preparations of the ovaries and spermathecae of a virgin queen (left) and after mating flight (right), made by Professor G. A. Kozhevnikov, a well-known Russian student of the anatomy of the honey-bee

Only after this does she lay an egg which she fastens or rather deposits on the cell base with her ovipositor.

The bees surrounding the queen in her progress over the combs form a cluster, or a retinue as it was called at the time the queen was supposed to be the sovereign ruler. Among the bees in the cluster facing the queen one can see young bees, just emerged from the cells, continuously touching the queen with their antennae, as if familiarizing themselves with her scent. As soon as they leave, their place is taken by others. Bees feeding the queen remain in the cluster for a longer time.

A closer observation will reveal that most of the bees of the retinue crowd round the queen's abdomen and that some of them lick it with their flexible tongues. Formerly it was believed that the bees were thus grooming the queen, removing every speck of dust from her body, but now C. G. Butler has established that, by licking the queen, bees take a secretion from her body. The instant a bee has taken from the queen's chitinous abdomen a drop of this invisible glandular secretion—the glands have not yet been fully studied—she leaves the retinue and rushes away, stopping now and then to share this drop with the bees she meets.

Before laying an egg the queen looks into a cell to see if it is clean; to deposit an egg, she inserts into the cell her abdomen. Here the queen is represented alone so as the better to see her, but actually she is surrounded by her retinue

The fertility of the queen largely depends on the state of the colony in which she was reared and on the number of nurse-bees that looked after her when she was a larva. The better the larva was fed the more ovarioles are developed in the queen's body, the greater is her egg-laying capacity.

But the real productivity of a queen is finally determined not by the conditions of her rearing and feeding: the quantity of eggs she lays depends also on the number of bees feeding her during the breeding period. When a colony has a sufficient number of nurse-bees and enough food for them, the queen is fed more and lays more eggs. If, however, for some reason or other the colony has few nurse-bees or the food stores are meagre, the queen's productivity is accordingly lower. Thus we see that under normal conditions the queen lays in proportion to the quantity of brood the colony is capable of rearing.

A queen's fertility, however, depends not only on the composition of her retinue and the food stores. Soviet apiarists have accurately timed the "working day" of the queen in an observation hive. Second by second they observed her behaviour; centimetre by centimetre they measured her progress over the combs.

Observations conducted for many days showed that to lay an egg in empty combs, the queen covered about a centimetre; the average time for laying an egg was half a minute. Having deposited about thirty eggs, the queen rests and receives food from the sur-

A healthy fertile queen can lay only if she is fed royal jelly by the workers. The more food she receives the more eggs she can lay

rounding bees. The time spent for rest has been calculated to be about six seconds per egg. Thus, the queen spends a little over forty seconds on laying one egg, on moving from cell to cell and on rest.

As there are twenty-four hours in a day and the queen takes food about one hundred times a day it is practically just because of lack of time that she does not lay more than 2,000 eggs.

These are, of course, average figures. Record queens in exceptional colonies can lay 3,000 eggs and more in twenty-four hours.

A laying queen moves spiral-wise, gradually increasing the radius of the laying area.

With a regularly-laying queen the space of the combs occupied by the brood is more or less circular, the circle being described clock-wise as a result of the slight movements of her body we have just spoken about. This concentrated laying saves the time of the queen, and since, as a rule, eggs are deposited on both surfaces of the comb it is easy for the colony to warm the brood.

In the course of time, however, when the combs are occupied partly by eggs and larvae and pupae developing from them and partly filled with honey and pollen brought by the adult bees, the queen spends more and more time in search of empty cells. Their number decreases from day to day and the queen has to walk longer distances between depositing eggs. Lack of space limits the queen's daily output first to a thousand, then to eight hundred, and later even to a lesser figure. But as the queen's laying decreases, her retinue also dwindles, and the nurse-bees feed her less than they would otherwise.

All these circumstances should be borne in mind in explaining the queen's ability (for a long time considered mysterious) to lay a very great number of eggs in spring, up to the main honey-flow, which the colony meets largely multiplied with a great force of foragers, and to decrease laying during the main flow.

Other causes, of course, come into play, such as the general state of the colony and its response to the conditions of the external surroundings. The result of all these causes is that in spring, when young bees emerge in a mass, the population of the colony increases and later, with decreased laying, begins to decrease.

Young workers develop from the eggs the queen deposits in worker cells. Drones emerge from eggs the queen lays (usually in spring) in larger cells. Exceptions to this rule have been noted, but all of them tend to confirm it.

A queen develops from an egg laid in a much bigger cell of peculiar shape, to construct which, a hundred times as much wax is needed as for a worker cell.

It has been discovered that the construction of the cell-cups from which queen-cells are built is connected with the bees that lick the queen's abdomen. If the queen is put in a cage through the walls of which the bees can reach their proboscides to lick her, life in the colony goes on undisturbed. But if the cage has double walls and the bees cannot touch the queen with their tongues, then a few hours after the queen has been encaged the first cup-cells are started. It goes without saying that under natural conditions this also happens—when the workers cannot obtain and pass to one another the secretion of the queen's integument, usually as a result of the queen's illness or death.

If for some reason or other a colony has lost its queen the workers can convert into an emergency queen cell any cell containing an egg or a sufficiently young larva. If the egg continued to develop in an ordinary worker cell a worker would emerge — one among the thousands of sexless labourers of the hive. Her entire six-week-long life (in summer), from the moment she emerged to the last beat of her little heart in the abdomen, would be spent in work on the combs and under the blue sky, flying from flower to flower. But an entirely different destiny is in store for the egg in the cell which the bees have enlarged to rear a queen.

All through her larval and pupal period the queen-to-be receives from the nurse-bees the piquant, slightly acid royal jelly given to all eggs during the first three days after laying, instead of the honey mixed with pollen fed to worker larvae beginning with the fourth day.

Minute analyses performed by numerous researchers one of whom analysed the royal jelly from 10,000 queen

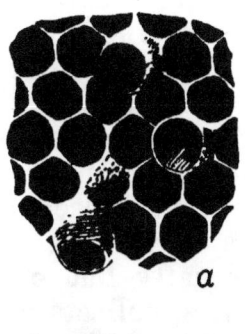

a

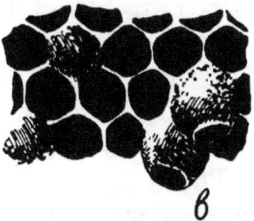

б

c

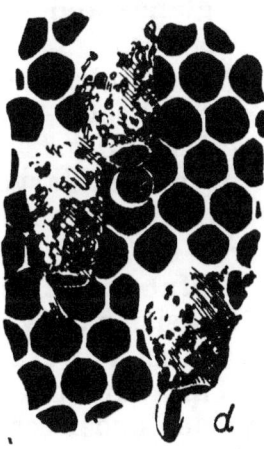

d

Bees usually start constructing queen cells by making a cup out of an ordinary worker cell (a). Then they draw the walls of the cup and make it a queen cell (b). When queen larvae have grown, the bees seal the cells (c). A grown queen bites through the capping of her cell from within and emerges, swinging it open like a lid (d)

cells, have shown that queen larvae receive more proteins and fats and less sugar than the worker larvae. Royal jelly, the study of which is in its initial stages, differs from usual brood food in other respects, too. Recently it was discovered to contain a large percentage of pantothenic acid and other vitamins not present in the food of worker larvae.

Great indeed are the changes this special food works in a larva: the queen differs from the workers not only externally but even more so functionally. She visits no flowers, collects not a drop of nectar, not a grain of pollen. She is capable of living three, four, even five years, almost fifty times as long as a worker born in summer, but she spends her life in the dark hive leaving it but two or three times. Day and night, spring and summer, she spends going over the combs laying eggs.

There is evidence enough to show that the transformation of a worker larva into a queen is brought about mainly by the change of diet.

When, for some reason, a colony has to rear a queen from a larva in a more advanced stage, which would otherwise soon be put on to common food—the mixture of honey and pollen—such a queen usually turns out to be inferior. However lavishly the larva may have been fed, the queen is born under-sized, and is hard to tell from a worker. Her life span, too, is shorter. Such queens have a smaller number of ovarioles in their ovaries and some of them have wax glands and pollen-baskets on their hind legs, just like worker-bees.

These intermediate forms, workers with queen properties or queens with worker properties, are now obtained experimentally.

If worker larvae between six and ninety-six hours after hatching are placed in queen cells and several queen larvae of the same age are put in worker cells, a sufficiently skilful experimenter will get two series of complementary transitional types ranging between a normal queen and an ordinary worker.

This explains why from a worker larva grows a bee capable under certain conditions of laying drone eggs, while a queen larva fed on "worker rations" develops into a queen with a number of worker characteristics.

Consequently, the eggs the queen honey-bee lays in similar cells produce larvae which develop, according to the way they are reared and fed, either into sterile workers or fertile females—queens.

How convincingly these examples demonstrate the decisive influence of food and environment on a living organism! How tangible this proof of the correctness of the fundamental premise of Michurinist biology!

As a rule, there is only one queen in a normal bee-colony.

The solicitude of the workers for the only perfect female of the colony is so consistent and persevering that it cannot fail to attract the attention of anyone observing life in the hive.

Even before the queen is born, while she is maturing in the wax cradle the bees surround her with unflinching solicitude. While the queen cell is unsealed, it is attended by a crowd of bees, now one then another of which pops in to feed the egg. When the queen cell has been sealed, the bees shelter it with their bodies. At the slightest lowering of the temperature they quickly move their legs, fluttering their wings and quake with their bodies, as though themselves chilled, to warm the queen cell.

The young queen returning from her mating flight is instantly surrounded by bees. We have described above how the queen is cared for while laying.

If by some disaster the colony is starving, the queen is fed as long as there is a single bee able to move about. The last act of the last bee dying of hunger is to give the queen the last drop of honey left in the hive.

Bees thus offer a visual and palpable demonstration of the usually hidden processes making heredity conservative, stable.

In analysing such processes, Academician T.D. Lysenko quotes in his works an example of two millet plants, one of which with its stalks, roots and panicles weighed about a kilogramme, the other less than a gramme. These plants grew under greatly different conditions, the former on very rich, the latter on exceedingly poor soil. One was a thousand times heavier than the other.

Yet the difference between the plants grown from the giant-millet seeds and those grown from the dwarf-millet was very small.

"Although the plants sharply differed in their development and departed in opposite directions from the norm, the processes upon which the continuation of the race depends most were fed quantitatively and qualitatively close to the norm. After all, the size of the seeds taken from these two plants—plants which exhibited a thousand-fold difference in weight—was almost exactly the same. Moreover, the embryos in these seeds, being the most important part, differed still less from each other. And, finally, the most essential parts of the embryos must have differed least of all."

All this is in accordance with natural laws. As Academician Lysenko says, there can be no equalization in the supplying of the food elements necessary for the different processes in a healthy organism. The most important processes and organs are the most rigorously protected both from lack and excess of food.

In animal organisms as well as in plants, the qualitative and quantitative *norm* is observed here in every way.

In bee life this principle of non-equalization is most apparent in the care for the queen.

Bees feed the laying queen only with the secretion of their brood-food glands. Consequently, however heterogeneous the food the bees have collected, the queen receives it radically changed in the form of royal jelly. Such is the nature of the attendant bees feeding the queen with royal jelly that they serve as a kind of living filter protecting the active vital centre of the colony. This filter is highly effective: if bees are fed honey coloured with some harmless substance, the mixture of honey and pollen given to larvae in an advanced stage is also coloured. But even in this case, royal jelly remains pure white, which proves that the colony uses nurse-bees to eliminate the external influences that might penetrate through various qualitatively different foods collected.

The nature of the colony is such that the nurse-bees which feed the queen and the larvae are carefully protected from direct external influences which might have a too pronounced and coarsening effect on the quantity and quality of the food they excrete to feed the brood.

It must be added here that the filter provided by the nurse-bees seems to be insufficient in the case of queen larvae.

If we bear in mind that, like all young organisms, and embryos in particular, larvae are extremely sensitive to external influences, we will not be surprised that queen larvae are given an opportunity of using the selective instinct which all living organisms possess and refusing unsuitable food. Perhaps this is why queen larvae are supplied with food in such abundance that some unconsumed food is always left in the cell after the emergence of a queen. True, mature virgins awaiting in their cells their time to come out are known to have tried to eat the dried food in the cell. This food, incidentally, can keep in excellent condition for years.

The fate of young worker and drone larvae is somewhat different. During the first three days after the laying, as long as the larvae are more sensitive to external factors, they, like queen larvae, are surrounded by a reinforced guard and a several-layer-strong blockade.

From the fourth day a layer of the blocking filters protecting worker larvae is removed and they are fed rougher food brought from without by a shorter road.

These details are very important in order to understand the biology of the bee-colony, its nature, its heredity and, to a certain extent, to understand why man's selective breeding has hitherto failed to produce a fully domesticated strain of bees.

Reproduction of species is the most important process in the life of an animal or plant organism, and bees protect and guard the queen in every way so long as she fulfils her mission and lays well.

The commotion and chaos that arise in a hive after the loss of the queen have been often described. Every bee-keeper knows that in such cases the bees begin to rush in panic óver the alighting-board and the outer walls of the hive.

Inside, the signs of alarm are still more apparent. For a long time, the bees keep running about and overtaking one another. The excretion of the queen's chitinous integument (a secretion not yet sufficiently studied but very important for the vitality of the colony) has ceased to take part in the colony's metabolism, and this noticeably changes the behaviour of a great number of bees. Some time after the disappearance of the queen, all the single running bees unite in a real stream. In hundreds the insects glide in a circle over the combs, driving along with light flaps of their quivering wings.

If there are no more eggs or young larvae in the hive from which a new queen can be reared the queenless colony may eventually perish. But if the queen has left eggs or young larvae, the bees of the queenless stock will quickly clear the space around the selected cells and begin to reconstruct and enlarge them, changing each of them into a cup-like foundation for an emergency, or post-constructed queen cell.

As soon as the reconstruction of the first cells into queen cells has been started, the excitement of the colony subsides, but life does not assume its normal course: until a new queen is born the bees are less diligent in foraging, build no new combs and are so irritable that it is better not to approach the hive unless serious need arises.

One can often tell by merely watching the bees at the entrance that the new queen has emerged. In the hive the very buzzing of the bees seems merrier.

It is no easy task for a colony to rear a new queen, yet bees are wary of strange queens reared in strange colonies; sometimes they absolutely refuse to accept one. An attempt to introduce a strange queen may therefore end in a failure. Bees are commonly held to recognize a strange queen by her smell, but probably there is more to it than just that. Bee-keepers resort to countless ruses to deceive the bees, but none can be wholly relied upon. Bees have been known to bite through sealed queen cells that the bee-keeper tried to introduce from strange hives.

All these well-known facts seem to prove that a bee-colony needs a queen of its own. Just as many plants from which the terminal bud has been removed start budding in several places, so does a colony deprived of its queen invariably build several new queen cells.

The young larvae chosen to become queens are fed royal diet and the life of the colony proceeds without any perceptible changes.

It sometimes happens that a colony is no longer satisfied with its queen (in many cases this doubtlessly depends on the secretion of her chitinous integument not being enough for all the bees of the colony) and the bees start building cup-cells, new queen cells in which to rear new females, one of which will supersede the old one.

The facts related, and many similar ones, prove the incorrectness of old opinions which held that the queen was "honoured, revered and venerated by all the bees," and was, if not a sovereign, at least the "head of the colony," or its "servant."

If analogies are necessary, the queen-bee is the living centre of the colony towards which flow streams of royal jelly and from which spread counter-streams of an excretion unifying the mass of bees in the colony. At the same time she may be likened to the cone of growth of plants: the queen is the growing point and the fruit bud of the colony.

ABOUT THE COLONY IN GENERAL
AND THE DRONE IN PARTICULAR

The Specialization of Organs and of Organisms. What Constitutes the Material Basis of the Organic Unity of Thousands of Bees Forming a Bee-Colony. The Factors Governing the "Mass Actions" of Bees in the Colony.

Under normal conditions worker-bees make up 99 per cent of a bee-colony, the remaining one per cent (if we disregard the only female) consisting of several hundreds of male bees—the drones—stout, ungainly, round-headed creatures much larger than the workers. The drone's strong wings produce a dull low buzzing as they quickly carry his thick body in flight. The sound may strike one as both businesslike and threatening, but both impressions are illusive, for the drone is quite harmless. He has no sting and his mouth parts are very delicate. While workers use their mouth parts for various jobs, the drone takes part in no work in the hive. He usually stays in the hive on the combs and leaves the hive only for a few hours when the day is at its hottest. In his orientation flights the drone may fly a considerable distance.

Once out of the hive, the drone seldom visits the sun-warmed flowers, from which the males of other bee species collect nectar and pollen, as did the male ancestors of the honey-bee in the past. The drone of today cannot do even that. His body is not adapted for gathering pollen and the structure of his mouth is exceedingly primitive: the short proboscis is capable only of taking honey from an open cell or from the tongue of a worker-bee.

The fact that the drone eats no other food than that gathered by the workers is very significant. Here again we come up against the filter system mentioned above with which the colony protects its reproductive centres. We see, however, that in the case of adult drones these filters are comparatively fewer: his diet is rougher and less highly processed than that of the queen. The drone can leave the hive at will, so that it would seem that even the nurse-bees are more carefully protected against external influences by the nature of the colony than the drones, whose direct function is reproduction.

This may suggest the question: Are not the characteristics of drone behaviour we have just described a refutation of the statement according to which the reproductive centres are specially protected by the bee-colony? But the fact of the matter is that a grown-up drone needs no such protection as the nurse-bees, to say nothing of the queen. The study of the process of development of the internal organs of the drone shows that the male bee is born with sex organs completely developed. His development is completed during the pupal stage under the wax seal while he is hidden in the brood chamber with its carefully regulated conditions. By the moment a drone leaves the hive for the first time, no unfavourable conditions are capable of materially changing the perfectly ripe cells in his organs.

The male bee's most striking feature is his goggled head. It is practically all eyes, consisting of 12,000 to 16,000 facets all merged into a single eye, and the ocelli, low in the forehead, below which are long antennae of 13 joints—one more than ˙in those of the worker. The number of nerve-cells in the drone's antennae is 30,000 — five times as many as in the worker's.

The drone's eyes, and especially his antennae, are his organs for finding a virgin queen, and his wings—organs for overtaking her.

Drones do not take food themselves—they are fed by workers. This fact shows that the males of the bee-colony, too, are rigorously protected from the influence of the external conditions

Why, one might ask, should drones living in the same nest as the female queen be provided with organs for finding a queen and following her? Why does he need to look for her? She is on the combs, by his side. But we may see that inside the hive drones will pass by the queen without taking any notice of her even if she has not had her mating flight.

When, however, the hour of that flight draws near, the lazy, languid drones on the combs become unrecognizable. They move faster and make for the entrance, and when the queen makes her exit they launch in pursuit of her. In an instant the drones take off the landing-board, mount into the air and with a loud buzzing crowd after the queen. The speed with which a drone flies is much greater than that of the worker.

It is believed that drones from strange hives flying about at the moment join the crowd of the virgin's "suitors."

The fact that the drone and the queen always come together on the wing has made biologists suppose that the absence of man-bred strains of bees is largely due to the difficulty (or impossibility as it was thought before) of selecting drones for mating with the queens.

Bee-keepers often see the flight of the queen and the crowd of drones after her but very few have seen the details of the mating flight, although it always takes place during a calm, sunny spell.

A few minutes after leaving the hive the queen returns to it already impregnated. She lands on the landing-board and rests for a few seconds, or proceeds into the hive at once. One after another, the drones come back, too.

It is not known which of the drones has mated with the queen, but what is known is that he is not and cannot be among the males returning from the marriage flight: the drone that has fulfilled his mission in life dies instantly. His successful pursuit of the queen ends in death for himself. The other drones may peacefully live through the summer.

It has been noted above that the drone sips honey from unsealed cells or receives it from the workers. This anatomic peculiarity makes him dependent on the worker-bees for sustenance and is the cause of his death. As soon as the best nectar-bearing plants have ceased blooming and the honey-flow is ended, the bees turn against the drones, feeding them less and less, which enfeebles them in a very short time. This may happen in summer, too, if the honey-flow has ceased owing to bad weather.

On an ill-starred day in late summer the free and easy life of the drones comes to an end: the stores of honey are completely sealed and the field-bees come home without bringing in fresh nectar. This is a signal for what has often been described as the "workers' riot" and their "revenge on the lazy drone." All the drones are driven from the hive.

If some bees and drones are taken from a hive and put into a glass jar in the season when trees and grain crops are in bloom it will be seen that, though removed from their natural surroundings, the workers will manifest a touching solicitude for the male members of the stock. If a drone stretches his proboscis to a passing worker she will share with him the last drop of honey, as if they were in their own hive and not in a jar.

Such idyllic good-will is at crying variance with what takes place in the same jar in autumn. The herbs are all out of bloom, the leaves have turned yellow and the seasonal changes occurring in nature have a most deplorable effect on the relations between the insects. No sooner are the workers put in the jar than they attack the drones and gnaw at their wings. Catching hold of the heavy drone with all her six legs, a bee will beat against the glass trying to fly towards the light with her load. Unable to hold the drone long, she drops him on the bottom of the jar and then again picks him up and again tries to get away with [him [in response to the instinctive urge to remove the drone from the jar, though there he is absolutely incapable of hurting the bees in any way.

Drones being driven from the hive. Observing dozens of worker-bees on the landing-board dragging away unwieldy drones, the bee-keeper knows that autumn is near

Here, under the artificially created conditions of a simple experiment, this blind intolerance towards the drones on the part of the autumn bees is a tangible proof of the automatic nature of instinct, a characteristic which we shall have many an occasion of observing later.

The summer is gone and bee-colonies start preparations for the winter. Merciless to the stingless drones, the bees drive them beyond the entrance which they will cross never more, since a living wall of guards bars the outcasts' way to the warm home; and when evening falls the drones perish from cold one after another. A cool night wind sweeps their light bodies and whirls them away together with the first dry leaves, the precursors of autumn.

Why, then, does a colony breed so many drones? Perhaps because, with a great number of them, a queen has more chances to mate in her first mating flight, and the sooner she starts laying eggs the stronger the colony will be.

We may also suppose that the denser the crowd of drones following the queen in her flight the better she is protected against all kinds of winged pirates. True, the cost of supporting such a number of body-guards is rather heavy, but to protect the queen is the colony's prime concern. Incidentally, do plants not produce more pollen than is necessary to fertilize the flowers? Nature is indeed unstinting where reproduction of the species is concerned.

Drones are not permanent but only temporary members of the colony. Under the conditions of the temperate zones of the Soviet Union there are no drones in the colony for the greater part of the year. As already mentioned, they appear in the hive at the end of spring and are expelled at the beginning of autumn.

All that has been said about them shows that drones, as well as queens and workers, cannot live isolated. An individual bee is incapable of living alone, cannot exist by itself: this is a generic characteristic of the honey-bee.

A queen-bee put on a comb filled with honey and bee-bread will soon perish if she is alone, if there are no hive-bees to feed, and warm her.

Neither can the drone, whose very name has become synonymous of a lazy person waxing fat on the work of others, live without a warm nest built and collected by the entire colony and containing stores of food.

Even the vigorous and active worker-bee cannot long exist alone: to become nourishing food, nectar must be turned into honey, but one bee cannot make honey. A bee has exceptionally developed wax-glands, but alone she cannot build a single cell, to say nothing of combs: one bee cannot build. She may exert all her efforts to warm herself outside the hive, but the first cold spell will chill her to death: one bee cannot protect herself from cold.

Apparently the anatomy and physiology of the honey-bee are adapted to life in a colony, to the life of the colony. Like any other product of natural selection, under normal conditions the bee-colony obeys certain laws in its growth: it breathes and consumes food. Like all living things, it obeys the laws of heredity, variability, and survival.

Although it consists of many thousands of individuals, the bee-colony is at the same time an entity—a parcelled, discrete natural community, a kind of "organism of organisms." Link and subordination between parts in this entity are so shaped and regulated that every individual taken singly is connected by innumerable criss-cross ties of dependence with all the other members of the community.

Collective feeding of the brood by the nurse-bees physiologically predetermines the colony's unity. The basis of this organic unity lies in the ever-wakeful nest with stores of collectively-obtained and prepared food in collectively-built combs,' where the temperature and humidity are also collectively regulated.

Practical bee-keepers have always considered the colony as a unit and manipulated it as such. But the biology, anatomy, physiology, and genetics of the bee-colony as a unit have been studied incomparably less than those of the individual bee.

That is why to this day it is not clear why an individual bee, or groups of bees, are able to perform just those actions which satisfy the constantly changing needs of the

Mutual feeding is a character of immense importance for the unity of the individuals making up a bee-colony. Each act of feeding is a link in the chain of "communal metabolism" making a unified whole out of the mass of bees

colony. For instance, how does the "order" to intensify the fanning of the hive during the time of great heat originate and how is it communicated to individual bees? It would seem that such an "order" is called forth by the sun-rays raising the temperature in the hive. But then why do not all the bees fulfil this "order"? Why are only some of the bees engaged in fanning? Besides, they do it differently on the alighting-board, at the entrance and in the brood chamber.

And this is one of the simplest examples. We have here a responsive action to a clearly perceptible change in a condition influencing the whole of the colony. But the colony consists of tens of thousands of bees, all of which are in relatively different states and, consequently, respond to the action differently. We may be led to think that one of the advantages a strong colony has over a weak one lies in there being a large number of bees in different physiolog-ical states, owing to which it can react to changing external conditions more accurately, more minutely and more precisely, thereby being more perfectly at one with the external conditions.

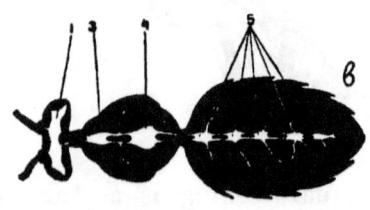

The nervous system of a worker-bee: a) side view and b) seen from above. The most important parts of the nervous system are: 1. the brain or the optic lobe in the head above the pharynx; 2. suboesophageal ganglion; 3. first body ganglion; 4. second body ganglion; 5. abdominal ganglia. The figures on page 169 show the difference in the structure of the brains of worker-bee, queen and drone

But if it is so, how do individual bees get the "report" that the colony has lost its queen, for instance?

How are the bees "informed" that the time has come to gather pollen from the hazel catkins on the border of some remote wood?

How is the "order" to drive out the drones given? Among the bees attacking the drones are many who do not leave the hive and, consequently, have no means of knowing that nectar-bearing plants have ceased to bloom.

Why do bees (not one bee, but groups of bees) begin to build queen cells?

Why do bees start enlarging cells in the combs for the queen to lay drone eggs a few days later?

Each of these actions is undoubtedly called forth by certain stimuli influencing the manifestation of various reflexes, instincts and reactions, and one who daily witnesses hundreds and thousands of bees performing in harmony a sequence of various operations cannot wave aside the questions: how the signals are transmitted, how do the respective responses appear, what unifies the isolated actions of the individuals.

The study of bee-colony biology suggests answers to these questions.

It has long been known that the bee is endowed with central, perypheral, and sympathetic nervous systems. Parallel to these, there is a specific system of connections uniting individual bees into a whole,—the colony. This system receives signals and responds to impulses, it directs the actions of the bees and unifies their activities. In this system, the nervous systems of the individual bees serve but as receptors and transmitters.

Later we shall dwell at some length on the initial links of this "wireless" nervous system of the bee-colony. But even now it is clear that, built largely to the image and likeness of organisms which are constantly perfected through selection, the bee-colony has developed within itself a specific unity of action and harmony of all vital processes. These processes in the bee-colony are, so to say, animated and personified, and resolved into the activities and behaviour of groups and individual insects.

The observation hive has enabled the bee-keeper to see the colony in its natural surroundings. The scientist studying the bee-colony can, by observing this physiological play-acting, penetrate into mysteries of living nature which are perhaps nowhere so open to the observant eye.

Unifying the thousands of bees composing it into a parcelled and at the same time *single* entity, the colony appears before the student as a specific biological unit, as a "divided indivisible," where the whole exists for each and each individual exists for the whole, where the whole and its parts are represented in a living unity.

THE WAX FOUNDATION

Construction and Location of Brood-Nest, Combs, and Cells. Origin of Bees' Building Instinct. I. P. Pavlov on Two Forms of Bee Behaviour. Comb Construction, "Water Supply" and "Sanitary Services" in the Hive.

If we tried to represent the interior of a modern movable-frame hive on a bigger, man-size, scale we would draw a town closed up in a gigantic cube and hanging over an empty level square of some 20 hectares.

A scant diffused light penetrating from one side allows here and there a glimpse of parallel rows of low-hanging fifty-storey constructions, reaching high up and lost to sight in the darkness above. The foundations and plinths are high over one's head, in the ceiling, which is the unseen base of the entire pendent town.

The buildings hanging over the square are all built according to a uniform plan, they are long and exceedingly narrow, as are the streets in the town, where two persons can hardly move side by side.

To the picture we have drawn we must add that each building from bottom to top—or rather from top to bottom—overlooks the two streets, to right and to left, with thousands of niches leading into long, low, thin-walled hexagonal cells, which in the enormous structures we have visualized could be built only of ferroconcrete.

But the interior of the bee city is built of wax produced by the bees themselves. The wax structures which, as someone has remarked, combine lightness with strength, and elegance with utility, are to this day an object of admiration for engineers and naturalists.

Today it is a known fact that bees secrete wax to build combs as the spider secretes the material to spin his web. Few know, however, that for twenty centuries—two thousand years!—ever since Aristotle's statement that bees gather wax from flowers, biologists could not find a key to its origin.

Under natural conditions bees make their homes in hollow trees, in fissures in cliffs, sometimes in the ground. Bee-keepers of old hived their bees in skeps in the south and in log hives in the north. The modern practice almost everywhere is to keep bees in hives made of wood with light wooden frames inside for the combs to be built on.

Wherever bees may live, the arrangement of the abode is roughly the same: combs attached to some horizontal surface hang vertically with narrow spaces between them, where the inhabitants of the hive are busy day and night.

There is a reason why bees build the combs downwards and not upwards. In attaching the combs to the ceiling of the cavity where they make their abode, the bees plaster it with propolis and wax, stopping every crack and fissure; warm air collects under the ceiling and the heat generated by the bees is preserved for the benefit of the colony.

The space between two neighbouring combs is from ten to twelve millimetres wide, twice the average height of a bee, which enables bees to move over the hanging combs back to back without touching.

A comb is a wax construction with a thin dividing midrib or septum; a standard frame contains about 7,500 hexagonal cells of nearly $1/4$ cubic centimetre each, situated back to back in fifty rows on both sides of the comb.

Up till recently it was so firmly believed that bees build with unerring precision, that 5.5 millimetres, the length of the cell diameter, was proposed as a standard of measure. But it has now been proved that combs are not absolutely exact either constructively or dimensionally. By measuring with extreme accuracy a very large number of cells it has been established that the angles of the prisms and the trihedral planes of the bottom forming the figure known in geometry as the Maraldi Pyramid, are perfect only in four cases out of a hundred, while in the ninety-six other cases the cells are non-standard.

Still, every detail of the wax construction is so perfectly, one might say cleverly, executed that Darwin was well justified in saying: "He must be a dull man who can examine the exquisite structure of a comb, so beautifully adapted to its end, without enthusiastic admiration."

Inside, a beehive contains several combs with tens of thousands of hexagonal prisms in parallel rows on both sides of the comb and each closed at its base by three diamond-shaped areas forming an inverted pyramid. The angles of the rhomboids are equal to the corresponding angles of the cells.

The strict geometrical proportions of the cells attracted

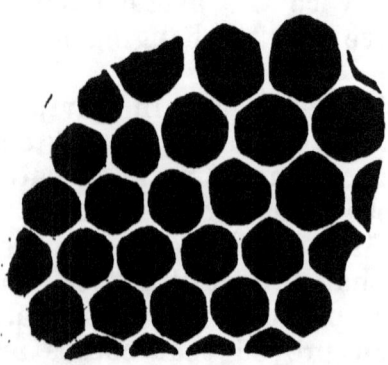

Normal drone and worker cells, and also transition cells in between

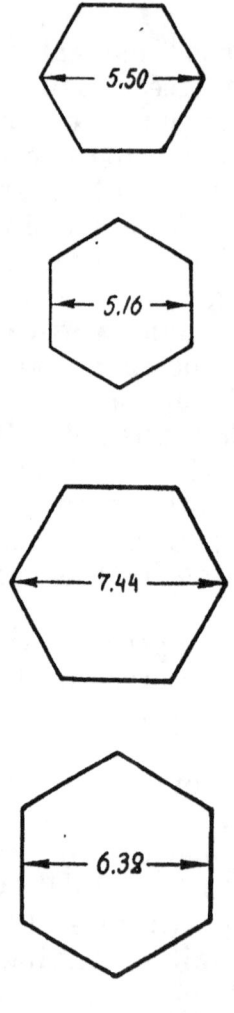

Average size of worker cells (above) and drone cells (below) in millimetres. Minimum and maximum sections of the hexagon are given

68

the attention of mathematicians long ago. After careful calculations, they arrived at the unanimous opinion that under existing circumstances bees have practically solved in the only possible way a recondite problem in stereometry, making their cells of the proper shape to hold the greatest possible amount of honey, with the least possible consumption of wax in their construction.

Calculations show also that this aim can be achieved, provided the acute angles of the three planes forming the base of each hexagon each measure 70°32'. And this is exactly the case with bee-cells.

The opinion according to which young bees learn how to build combs by observing their older sisters at work has long been discarded. In an experiment in which all old bees were removed from a hive, the young bees started building combs soon after emerging from their cells without anybody to teach them.

Like every other inborn action, the building art of bees is a blind, automatic and uniform repetition of the experience of past generations. It is an instinct, in obeying which the unconscious actions of individuals manifest the historically developed experience of the whole species, an experience which has become a law of life.

But when domesticated bees, after nicely drawing a full sheet of factory-made comb foundation, start reconstructing worker cells into drone cells, or when they build wax columns to support crumbled combs, and in many other similar cases, the colony manifests something that looks very much like an understanding of its needs and an ability to adapt itself to existing conditions.

When a difficulty arises, as, for example, when two combs meet at some angle or other, the bees build and rebuild one and the same cell in many different ways, sometimes returning to the shape they have previously discarded. Such actions on the part of the bees bear a strong resemblance to experimenting, but we must not be misled into the idea that bees possess reason.

In his time Engels noted that: "... a planned mode of action exists in embryo wherever protoplasm, living albumen, exists and reacts, that is, carries out definite, even if extremely simple, movements as a result of definite external stimuli."

We must, therefore, always remember that under certain conditions even simple reflexes, to say nothing of instincts, may look very much like conscious acts.

Does not the work of the salivary glands seem "wise" to us when they secrete an abundance of saliva if the food is dry and needs moistening, and only a little if the food is liquid? Moreover, they lubricate with mucous saliva food that is to be swallowed and secrete thin, watery saliva if the mouth is to be laved. An instinct, on the other hand, is a much more complex, a higher and more *active* reaction of the organism to surrounding conditions.

It was bees that Academician Ivan Pavlov had in mind when he said that two types of behaviour, "a higher and a lower, an individual one and one pertaining to the species" were observable in insects.

These two types of behaviour are especially hard to distinguish in a social insect like the honey-bee. The lower activity pertaining to the species, which Pavlov described as "stereotyped, inborn, so-called, instinctive" is very closely connected in bees with the "activity based on individual experience." The latter, however, must not be identified with manifestations of reason.

K. Marx gave an exhaustive explanation of the difference between instinct and reason when he said that: ". . . a bee puts to shame many an architect in the construction of her cells. But what distinguishes the worst architect from the best of bees is this, that the architect raises his structure in imagination before he erects it in reality."

It is planning, proof of abstract thinking and of the ability to form abstract ideas, to investigate their nature, that distinguishes the activity of man from the actions of animals and from the behaviour of insects.

Like other insects, bees manifest in their building activities not reason but only instinct.

Comb construction may proceed at a very high speed—as many as several thousand cells in twenty-four hours.

Bees hang in live festoons like electrified chains from the top bars of the frames, parallel to the surface of the comb they are building.

The bee that is uppermost in each of the live chains clutches at the top bar with her fore legs while with her hind legs she holds the fore legs of the bee next to her, the latter doing the same for the next bee, and so on, to the end of the chain reaching sometimes down to the bottom. The separate chains are connected between one another by the middle legs of the building bees, so that a living pulsating fabric is the result.

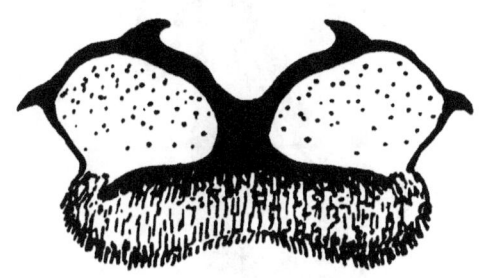

Abdominal segment of worker-bee showing wax mirrors

The honey-stomachs of the bees are filled with honey from the colony's store and chemical processes are taking place in the tiny laboratories of their bodies. Here carbohydrates are transformed into complex ethers, fatty acids, and saturated hydrocarbons. Honey is converted into building material, about four kilogrammes of honey being necessary for bees to produce one kilogramme of wax.

The festoons of comb-builders include both young, two-day-old bees and old bees up to the age of forty days, which may make up half of the cluster. But such bees with wax glands either undeveloped or atrophied take no part in secreting wax or building combs; they only help the builders to maintain the temperature in the cluster. Wax is secreted and the building is done by the bees whose wax glands are at the best stage of development.

A bee beginning to secrete wax leaves the cluster, runs upwards and pricking each wax scale with the spines on her hind legs passes them to the mandibles to masticate before affixing them to the ceiling of the hollow in the tree or the top bar of the frame. In this manner, the basis of the combs, which will soon hang over the hive bottom, is built.

After the first wax scale has been attached, the bee adds to it more and more, until she uses up all the scales from her eight wax pockets, when she returns to the festoon of builders. Her place is taken by another builder who continues the work started by the first bee.

A wax scale pierced by the spines of the hind leg is thus removed from the wax-pocket between the mirror and the overlapping part of the preceding sternum. The scale is grasped by the fore leg and passed to the mandibles. Masticated and softened, the wax scale is added to the comb being built

One bee after another adds pieces of soft, sticky masticated wax to the comb hanging like a spongy, porous mass over the empty space. Over them other bees are engaged in building cell foundations, from which a third group will start drawing cell walls.

A similar process is going on at other points a few inches apart.

This seemingly chaotic mass of hundreds or even thousands of bees is busy affixing, thinning and perfecting the wax texture of the comb which grows in length and in thickness and slowly solidifies.

Irregularities in comb construction may be observed here and there, but in one respect the bees hardly ever go wrong: their combs are, as a rule, strictly vertical; research has shown that the direction in which combs hang is determined by gravity.

Completed combs reach down to the bottom board, leaving space enough for bees to pass under them.

Comb construction proceeds on both sides, even the cell base being worked on from both sides simultaneously.

Dozens of building groups may be busy in a frame hive, and all their separate actions finally result in converting the space within the four walls into a standard brood chamber with five square metres of comb surface accommodating tens of thousands of bees.

It has been noted that combs, as a rule, are built only at the time when the hive receives a fresh supply of nectar and pollen. If a colony loses its queen, building will be discontinued even at the time of a heavy honey-flow. But with a well-laying queen and ample loads of nectar and pollen brought in by the field-bees, building operations will be carried on even where there is no room for more combs; the bees will satisfy their instinct for building by renovating old combs, dark with age, covering them with fresh light-coloured wax.

Bees, of course, do not build combs because they are afraid of a lack of storage for the nectar they gather. The development of the wax-glands which awakens the building instinct and urges the bees to build combs is physiologically connected with the appearance of fresh nectar and pollen.

Wax is secreted *involuntarily* when fresh food is brought to the hive and the bees are busy storing and processing it: the more plentiful the supply of food, both nectar and pollen, the more abundant the wax secretion.

At the same time it is quite true that queenless colonies are very slack builders. This, of course, is also quite natural, for in bee physiology wax secretion depends not only on the quantity of food the colony receives but also on brood-rearing: colonies containing active nurse-bees produce more wax, other conditions being equal, than colonies in which there is no brood to rear.

To finish with this problem, we must mention inverse dependences ascertained here. It is a known fact that when a colony has no storage space available, bees forage less actively. In colonies where no wax is secreted and no brood fed, the field activity likewise considerably slackens.

This feature of bee behaviour, like those described earlier, is determined in the final analysis by physiological causes.

The "building talent" of bees is not limited to constructing new or repairing old combs consisting of cells. As summer draws to a close, bee colonies start preparations for winter, beginning with the thorough caulking of the ceiling which is plastered with a strong bee-glue — propolis. Bees also use propolis to stop all cracks and crevices in the hive and to reduce the entrance leaving only a narrow slit for the winter.

If you see a bee not in a flower but on a leaf-bud, you may be sure she is collecting bee-glue, propolis

The propolis used by bees for all such purposes is gathered from the buds of certain trees, usually in the hot hours of the day when the sun has warmed the tacky substance and rendered it plastic. The propolis collectors tear with their mandibles and fore legs chunk after chunk of the gluey material, then masticate each chunk and deposit it in the pollen-baskets. This is a time-consuming operation, and a worker may spend an hour collecting a single load of propolis.

When the propolis-collector reaches the hive, she does not unload herself as pollen-gatherers do. She makes her way to the top bar of a frame and waits there until the hive bees need glue. They come to the waiting bee and remove bits of propolis from her legs with their mandibles. It is hard work and both bees take a firm hold on the frame surface and pull with all their might. Having obtained a little piece of propolis, the builder carries it to where the work is in progress and puts it to immediate use.

You see what an effort the bee must make to tear away particles of propolis from the load

All the time the unloading is going on (and it may take several hours) the propolis-collector waits patiently, now and then stretching her proboscis to passing bees, asking them for food. At last the entire load is removed from her legs and the bee is free to move.

If it is still hot outside, the bee will make another journey to bring more propolis with which she will again stay somewhere at top of the frames.

Bees are not very particular as regards the source of propolis and are known to introduce into the hive glue of quite unexpected origin. Propolis of a startlingly azure colour was once seen. It proved to have been collected from a nearby fence recently painted blue.

It would seem that collecting propolis is as much a vital necessity as comb building. Many bee-keepers have observed hundreds of bees gathering in front of the entrance on the landing-board at the end of summer, and, standing in rows with their heads lowered, performing a rhythmical movement, as if gnawing at the fibres of the board. Formerly these movements were interpreted as "harvest-home" celebrations. It is now established that the "gnawing scene" is connected with the gathering of propolis, which is so essential for preparing the hive for the winter.

The builders also reinforce the cell walls with a thin layer of propolis, as a result of which the combs, white when freshly built, soon become yellowish. As time passes they become brown and even black, owing to the accumulation of cocoons and cast-off larval and pupal skins and the remains of unconsumed food. This, too, accounts for the thickening of the cell walls, and particularly, of the bases, which results in the reduction of cell size. The cells would become too small for brood-rearing but for the bees gnawing off the cast skins and removing them from the hive. It has been observed that the older the comb the more thoroughly the bees clean and polish the cells.

Stopping all cracks and crevices is important for maintaining a constant temperature in the hive.

All these characteristics of bee behaviour are of importance, too, because they substantially influence the stability and uniformity of conditions within the colony, thereby preserving and strengthening heredity.

The arrangement of the brood chamber also manifests an inborn characteristic, a hereditary property of the honey-bee.

A standard hive with 12 movable frames and a super contains over 100,000 cells. The disorder in the hive could become simply catastrophic were it not for the fact that each part is used for a specific purpose.

The better ventilated part of the hive, the lower parts of the combs and the combs nearer the entrance, accommodate brood cells containing eggs, larvae and pupae. This part of the hive is enclosed within a ring of cells containing beebread, of which there may be thousands. Combs for the storage of honey—the most precious possession of bees, so attractive to their enemies—are farther removed from the entrance, nearer the walls and the ceiling.

All cells, and this can be best seen in the upper cells, are built not horizontally but at a slight slope, the angle of which is not large but enough to prevent honey flowing out of the cells. The uppermost row in a comb consists of pentagonal cells which, as building experts agree, makes the fixing of the comb to the top bar of the frame easier and more reliable. These pentagonal cells are filled with honey.

The interior disposition of the hive makes for warmth, too. In the centre occupied by brood the temperature is maintained at the level required for the normal development of eggs, larvae and pupae while the temperature in the outer areas is usually lower. The bigger a colony and the more space it occupies the greater the difference between the temperatures. This causes a circulation of the air, which fanning bees utilize for ventilating the hive.

Neither old-fashioned log hives nor skeps are to be seen today at collective-farm apiaries which are stocked with modern movable-frame hives

It is also an important feature that the combs filled with honey are located in the outer parts of the hive and in the upper, back and, partly, front areas of combs in the centre. In this way the brood chamber proper occupies a spheric space surrounded with honey on all sides, like the body of an animal covered with fat. Here the honey, a low heat conductor, constitutes a kind of barrier against fluctuation of outside temperature.

This plan is convenient for man, too. If the bees stored honey in a haphazard way, among the brood and beebread cells, honey extraction would involve the destruction of a colony, while we can take away honey from bees (especially in movable frame hives) without disturbing the normal course of the colony's life. The bee-keeper leaves in the hive the combs containing honey, beebread and brood, and removes combs filled with honey, replacing them with empty ones.

The picture of life in the waxen city would be incomplete without a description of its "communal services."

While in the centre of the hive the queen lays eggs in cells specially prepared for this purpose, while the nurse-bees feed the brood with bee-milk and foragers pass their loads to home bees, while pollen-gatherers deposit it into store cells, the winged guards at the entrance are keenly watching lest strangers or robbers enter the hive.

Water-carriers come with their crops filled with water which is used in the hive by nurse-bees to drink and to thin down honey in the processing of larval food. Ripe honey contains about four-fifths of sugar while in the honey used for brood food the sugar content is lowered to two thirds by adding about one sixth water.

To carry into the hive one hundred grammes of water, three thousand flights to the watering place and back are needed. Water is brought mostly in spring when an average colony consumes 1.5 to 2 litres, and even more, in a week.

During a heavy honey-flow no water is usually brought to the hive, the water evaporated from the nectar in the combs being sufficient for the colony's needs. Part of the water used by the colony can, under many conditions, be condensed in the hive owing to the difference of the temperatures outside and inside the hive. All water is supplied by condensation in winter.

Sometimes water is used in a hive to moisten the air. On dry days droplets of water are

The normal development of the eggs and larvae is possible only at a definite humidity of the air. Droplets of water stuck to the upper cell-sides in dry weather moisten the air in the cells of young brood

attached to the upper sides of brood cells for this purpose.

Bees are water-loving creatures and suffer severely if for some time they cannot get any water.

In spring or summer after several days of bad weather, no sooner does the sky clear and the sun appear from behind the clouds than thousands of bees may be observed leaving their hives and scattering over the humid earth, over the wet stalks and blades of grass and avidly stretching their proboscides to the sparkling drops of water.

Water-carriers begin supplying the hive with water in early spring

It should be added here that it is not with the help of their sight that water-carriers find water. Bees blinded by dark varnish painted over their eyes stretch out their proboscides towards water just as seeing bees do. Seeing bees never attempt to take in such liquids as oil or spirit, outwardly undistinguishable from water, while both blinded and seeing bees know a moist lump of earth from a dry one. All observations in this field lead us to suppose that bees find water by the humidity of air, which they sense with their feelers.

The activities of water-carriers and the "reservoir-bees" are closely connected with those of the fanning bees: in hot seasons or in the hot hours of the day this group of bees reduces the temperature in the hive to the needed level. At the time nectar is ripened into honey the air saturated with evaporated moisture is removed from the hive, for bees suffer from excessive humidity as they do from drought. In the bee-spaces, bees may be observed fanning with their outstretched wings the open cells with brood or nectar.

Other fanners are posted in serried ranks along the bottom-board up to the landing board. Their heads all turned in the same direction, they operate their uncoupled wings vigorously, as if flying on the spot.

If the bottom of an observatory hive is warmed by an electric bulb, for instance, the bees covering the brood will be seen to move over the combs from the centre to the periphery. Soon fanners will appear along the combs and the bottom-board, and the longer the hive is warmed the closer the ranks and the greater the number of fanners.

In about an hour after the lamp has been switched off the ranks of the fanning bees will thin out, and by and by the bees will come to the centre and resume warming the brood.

The sound produced by the movement of the thousands of tiny wings is not unlike the hum of an engine, with intermittent rises and falls, a sign of the rhythmic, harmonious work of the wings. Their simultaneous beatings set up air currents which combine into a strong ingoing flow.

Let us take a bee-smoker from the funnel of which escapes a thin puff of smoke changing its direction with the gentlest breath of wind. If we place the smoker at one end of the hive entrance we shall observe the smoke carried away from the hive as if blown from within, while at the other end of the entrance it will be sucked *into* the hive.

The melting point of wax is 62°C., and the cool air pumped by bee ventilation into the hive prevents it from softening.

On hot days a too narrow entrance may prove the cause of grave disasters, and if it is closed completely, but for a short time, a strong colony may perish.

The wax city has its own scavengers which clean the combs, the cells and the bottom of the hive, removing dead bees, drones' evacuations, and all kind of debris.

The body of a mouse which has been stung dead and which the bees are unable to remove from the hive, is varnished over with air-tight propolis; the same fate awaits a snail that has managed to get into a hive. If a pungent-smelling camphor-ball is put into a hive it, too, will be covered over with propolis, and so the air in a hive of a sound colony is always pure and sweet.

ADVANTAGES OF A SETTLED WAY OF LIFE

The Burrow of the Osmia Papaveris. The Ceratina's
Nest of Linear Cells. Bees in Snails' Shells. The Wasps'
Paper Spheres. First Wax Constructions. The Nest and
Heredity. The Incarnation of Motherly Solicitude for
the Offspring.

We shall leave for a while the bee community with its
bubbling life and watch a dark small bee—or wasp—in
a far-off corner of the garden busy at the side of a well-
beaten path. If we choose our position well we may see at once
that, burrowing into the sandy soil with all her six legs,
the insect is deepening a tiny hole, in which first its head
and then the thorax disappears, while light grains of sand
fly from under the quickly-moving legs.

Now and then the bee stops working and gets out of
the burrow crabwise, carrying a bigger grain of earth between her mandibles. In a moment she is back in the hole
and the sand is flying up again.

As the depth of the burrow increases the bee gets out
more and more often and not always crabwise removing
heavy pellets of earth and grains of sand. But while the
little builder is invisible the movement of the sand around
the black entrance shows that she is there and working very
hard.

At last the insect leaves the burrow and starts grooming herself, removing dust from her body with her hairy
legs, cleaning her eyes and antennae, which takes quite a
time, then she mounts into the air and, after making several circles and loops over the burrow, flies away.

And now the observers will see the most unexpected
things in the chain of events occurring before their eyes.

The winged digger soon returns pressing a roll of something crimson against her body with her legs. She takes
it down into the hole and a little later goes off for another roll.

If during the bee's absence we place a blade of grass
over the entrance, this sudden obstacle will make her drop
the roll and drag the grass away to clear the entrance. While
she is so engaged we can pick up and study the crimson
roll. It is a circular cut from a satin-smooth petal of wild

red poppy. With these discs of poppy petals the bee covers the bottom and sides of the burrow up to the very top. This is how the Osmia Papaveris species build their nests for depositing eggs in.

Many an entomologist has tried to find out why the Osmia uses poppy petals to deck her nest, and why the red poppy and no other. The supposition that the petals prevent the development of mould and fungi in the cell has not been finally confirmed. But the most astonishing thing is that the bee uses only the petals of the poppy and no other part of the plant.

No sooner has the leaf-cutter completed lining the cell with poppy petals than she ceases flying to the poppy and starts looking for cornflowers. From them she carries load after load of yellow pollen and a farinaceous mould is formed on the bottom of the poppy-decked cell. When the mould is big enough for the larva to feed upon, the bee covers the pollen with a layer of honey also taken from the cornflower.

The provisioning of the cell over, the Osmia lays the egg for which all this structure was built. But that is not the end: to provide for the safety of her offspring, the Osmia gathers the ends of the poppy petals and glues them together in a kind of canopy over the cradle in which her egg rests.

And then comes the closing act of the performance: the Osmia runs several times on the little crater round the entrance and scatters on the canopy the sand she has removed from the hole. Then she levels the ground, disguising her nest so thoroughly that the entrance becomes quite unnoticeable.

Having completed her labours, the mother-Osmia gives herself a last cleaning and flies away, leaving behind all her care about the nest she has built with such pains and the egg laid in it. She starts digging a new burrow somewhere else.

While the old Osmia is building a new nest, lining it with the satin petals and carrying into it food from the cornflower, a larva hatches in the first nest. It consumes the food its mother has provided for it and when all is eaten up, spins for itself a silky cocoon and goes to sleep a pupa.

After the required number of days have elapsed a mature Osmia awakens in the cocoon; she digs her way out of the burrow to light and warmth, and flies away.

Young female Osmias meet in the air young males emerging from other cells. After the mating flight female Osmias look for sides of roads, for soft, bright-red poppy petals unfolding in the morning, and for blue cornflowers with their picturesque yellow-headed stamens and prolific nectar-glands hidden among the bases of the dentate petals. . . .

The more scientists study the life cycle of the little Osmia, the more they discover in the structure of her body and in her habits various adaptive features connected with the burrow-nest in which she lays her eggs. Among the numberless morphological and neuro-physiological characters showing the link between the living Osmia and her inanimate nest are the structure of the legs enabling her to burrow in the earth, to collect food, to roll the petals and to press the roll against the body in flight; the structure of the mandibles capable of cutting circles out of poppy petals and of opening cornflower anthers; the structure of the proboscis capable of sucking nectar out of the flowers and moistening with honey the pollen in the nest; the instinct leading her unerringly from the field to the place where she is making her nest; and, finally, one more instinct making her forget a finished nest and start building another.

The Osmia makes several burrows in various places during her lifetime and for this reason cannot be considered an insect with a settled way of life. But then she does not lead a purely nomadic life characteristic of many crawling and flying species.

Certain bourgeois scientists engaged in the study of evolution hold that it was the wings that contributed most to the development of species. In the words of an entomologist: "Only the domination in the air made bees the primate of the insect world." Such a statement, of course, has nothing scientific in it. Why should domination in the air be mentioned at all?

Legions of winged species have not held and do not hold any dominant position among insects, while it is well known that the wingless ants have an equal right with the bees to be considered "the primate of the Hexapoda."

If it were necessary to name some particular factors as specially conducive to the many-sided development and complication of individuals in the animal kingdom, the nest should be mentioned first. Its appearance called forth the development of the building abilities of the species, the rapid development of spacial orientation, and of the ability to store food for the offspring.

All this is graphically illustrated by the study of the evolution of the Apidae.

The building activities of the Osmia Papaveris, like those of the Megachile, who decks the cradle of her young with rose or briar petals, and hundreds of other species of Apidae not so fastidious about the material they use, are limited to constructing a simple solitary cell.

There are hundreds of other species which construct more complex nests of cells situated in a line, one after another. That is how the elegant blue-black Ceratina, a tiny bee not more than half a centimetre long, builds her nest. She chooses a vertical stalk of a bramble and starts making a long narrow tunnel within it. She shapes the soft core with her mandibles into small balls and, pressing them against her body with her legs, crawls out backwards and throws the balls away. Putting one's ear against the stalk the little bee is busy in, one can distinctly hear the scratching sound she makes in the narrow passage.

It is worth mentioning that the Ceratina never makes her nest in the tops of long stalks, too far from the ground. She will not choose such a stalk even if its cross-section is large enough. The reason seems to be that long stalks are apt to break. But then the little bee will not build too close to the ground either: in rainy weather dampness may penetrate and damage the nest.

When the tunnel has been dug to the required length the little bee stamps the bottom hard, then lays an egg, puts some food over it and builds a partition which serves as the bottom for the next similarly constructed cell. A Ceratina nest may contain a score or more cells, but what-

ever their number, a "sentry-box" is invariably constructed on the uppermost partition, somewhat thicker than all
the rest. This "box" is bigger in size than the cells and in
it the mother-Ceratina stays at night with her head blocking the entrance of the tunnel, guarding her offspring from
enemies. She is old now and of a greenish colour.

Some time later the Ceratina's brood matures and starts
emerging from the cells, the first little bee coming out of
the egg laid last, that is, from the uppermost cell. Her
vacating the cell opens the way for her sister who is just
a little older, then the third, still older, sister leaves her
cradle, and so on and on, until the eldest sister, who has
had to wait longest, sees the light of day.

The old Ceratina still spends wakeful nights in her
"sentry-box" and will live here the rest of her life after
the last of her daughters has left the nest. After mating
flights the young bees will look for bramble or raspberry
stalks broken not far from the ground, dig tunnels with
their mandibles in the core of the plants, construct
cells, collect food, build partitions, and watch over the
brood. . . .

The bee-world abounds in countless varieties of the
simple burrow of the Osmia and the Megachile and the
linear nest of the Ceratina.

The Colletes, a species of solitary bees, glaze their
cells from within, covering them with a coat of lacquer-
like substance. Many social bee species whose males do
not assist in nest construction either also glaze their cells.
The Chalicodomes, or mason-bees, make their abode on
rock and cover it with an iron-strong shell made of sand
glued with their saliva. To smash this shell, a hammer or
some other tool is needed. The Tenthredinidae or carpen-
ter-bees, excavate tunnels in timber to make their nests,
and the Osmia tridentata bores hers inside dry raspberry
stalks. Some species choose living stalks, others dead ones
left over from last year and standing upright, still others
need hollow rushes used in the country for thatching roofs.
The Xylocopa and Lithurgus hollow out sound timber
while the Anthophora uses weathered timber. There are

species that make their brood chambers only in galls on leaves, in acorns, and such-like.

The Megachile centuncularis, or leaf-cutter, sews a little thimble-like sack. Using her mandibles as scissors, she cuts perfectly circular patches out of leaves and covers with them the bottom and lid of the nest; the sides of the tubular structure are covered with oval-shaped cuts of leaves. Different varieties of the Megachile use for this purpose the leaves of different trees.

Felt made of plant-down serves as material for the home of the black-striped red-spotted Anthidium or carder-bee. Another species of the Anthidium build cells of resin from coniferous trees and then collect them in a ball covered with a rind of resin and sand.

Different bee species build differently-shaped cells— round, angular, with an even or ribbed bottom, broad or elongated, horizontal or vertical, one- or two-sided, collected in open or closed nests of various types and constructions.

The burrowing bee species build in clay walls, in steep cliffs, often constructing sloping roofs over their nests. Others do not need all this and thrive in lime seams between bricks in mason-work. Some species choose bare well-beaten ground where they use cracks in the soil, and others are unable to start building anywhere but in soft turf.

Some species attach their nests and cells to branches, to rocks or under the roofs, while others require natural or artificially-made hollows and cavities. The Osmia bicolor uses an empty snail's shell in which she builds cells with partitions made of masticated leaves and then in an amazingly short time camouflages this wonderful nest with pine-needles, dry leaves, bits of straw, seemingly heaped haphazardly but actually glued so that the wind cannot destroy the tupee-like shelter.

There is one more type of bee building in shells of various slugs: the Osmia fossoria. Having built the cells and sealed the last of them, she starts digging a biggish hole in the sand to the depth of from six to seven centimetres. Unlike the burrow of the Osmia Papaveris, this burrow is not vertical but slopes at an angle of about 30°. Then the Osmia rolls the shell with its sealed cells into the burrow.

She rolls it gently towards herself, like a barrel, then covers the hole with sand, and levels everything around, so that no trace is left.

A naturalist who studied the camouflages of the Osmia varieties some half a century ago was fairly astonished at discovering that shell-building Osmias die immediately after exhausting all their energy on building their nests. So none of these insects live to see whether their offspring is preserved or which of the different ways of camouflage is the best for the preservation of the species; still, each uses a definite method. "How, then, do such instincts arise? Shall we ever be able to solve this most puzzling riddle of Nature?" the naturalist exclaimed in wonder.

We should very much like to start solving these problems right now, but we shall have an opportunity to return to them in the near future. For the present we shall content ourselves with stating that no riddle of Nature, however simple, can be solved easily and that all, even the most mysterious, can be unravelled by persevering students. Having said thus much, let us return to the survey of nest types built by the honey-bee's near and remote relations.

Without stopping to consider numerous other solitary bee species with nests of different types constructed in different places, such as a single cell, a linear row of cells, separate branching cells or a collection of cells, without speaking about the Halictus which builds clay combs to accommodate a community very much like that of the social bees with the only difference that it is a temporary colony, we shall treat here of the social wasps whose nests are covered with a cardboard-like substance of their own making. Wasps cover rows of one-sided combs consisting of standard-size cells with their openings downwards and resting on columns with several layers of paper made of tree bark and branches cut with their mandibles, masticated and processed with saliva. The cells contain food stores, eggs, larvae, pupae and young wasps about to emerge.

Then there is the entirely new world of the bumble-bees whose nest contains, heaped together, the first cells made of wax. The round cells of the combs are used for brood-rearing, and honey is stored in "honey jars" made of dark, brown wax. The bumble-bees' home is a comfortable wax dwelling. When a nest of the bumble-bees is built in a soil not loose enough, an old bumble-bee may be observed at sunrise standing at the entrance and buzzing its wings for a considerable time. The idea current in former days was that this was a bugler playing reveille for the community, but now we know better: the insect is ventilating the nest.

One more remark: there are no two species among the hundreds of the bumble-bee and social wasp species, as well as among the thousands of the bee species, that build their nests alike. At the same time the nests within a bee species differ but slightly.

Specific characteristics manifest in type and arrangement of nest and in the building process are often as definitely pronounced as anatomic or physiological characteristics, and the seemingly inanimate nest viewed from this standpoint becomes part of an animate species, a reflexion of its needs and nature.

Thanks to the knowledge of the laws governing the correlation of various parts of organisms, biologists can now reconstruct a skeleton from a bone of an extinct animal and form a pretty accurate idea of its mode of life. As a fragment of a vertebra reflects the whole skeleton and the specific characteristic, the details of nest construction reflect the hereditary characteristics of the insect that built it.

But it is not the nature of a species alone that is indicated by such tokens as poppy leaves on the cell walls and a pellet of cornflower pollen in an Osmia burrow, or the height on which is situated the base of the lowest cell in a bramble stalk and the "watch-box" over the uppermost cell in a Ceratina nest, or any of the spacial, typical and constructive features of the nest. The example of the honey-bee reveals one more biological aspect of the importance a fixed type of nest has for insects.

The eggs of various bee species, even considerably enlarged, look almost alike. Outwardly, the eggs of the Osmia Papaveris, the Ceratina, the Megachile, the wasp, and the honey-bee are scarcely distinguishable, but the insects developing from these eggs are widely different. This is explained not only by the differences in the structure of the egg matter and the embryo bodies, but also by the fact that these eggs can be transformed into insects only within definite brood-nests, strictly specific for each variety, where they are laid and where the conditions are such as to direct the development of the embryos along the natural course peculiar to each particular species.

It is well known that the endosperm of a seed, the cotyledons of a grain, the white of a bird's egg, the milk of a mammal are mentors, the directing media of the shoot, the fledgling and the young animal, all of which assimilate directing influences in a concentrated form from the food they consume. Thus we see that nests, too, strengthen the conservatism of the specific heredity and direct the development of an individual along a narrow path.

We know that each of the three castes composing a honey-bee colony can, under natural conditions, be engendered, develop, become a perfect insect and complete its life cycle only within a colony. In spite of the apparent independence of each individual, neither the workers, nor the queen bees, nor the drones can live separately by themselves for a period of any length, they are unable to exist one without the others. And now we should stress that colonies consisting of a sufficient number of workers and drones and possessing a well-developed and perfectly sound queen can live, multiply and develop only in a nest, on the combs.

The combs are built of inanimate wax but no bee-colony can live apart from them.

It is in the wax cells of the combs that bee brood develops. It is only on the combs, within a nest, that bees are able to maintain the temperature necessary for the development of the brood. Neither a worker, nor a queen, nor a drone can be born outside the cells. If there are no cells, a queen will not deposit a single egg. Bees that have no combs do not collect either nectar or pollen. And it is only in wax cells that nectar ripens into honey and pollen becomes beebread.

Isolated from their combs, bees will either build new ones or perish.

A bee-colony put in any suitable place such as a hollow tree, an empty box, a clay tube, or a skep, will immediately start filling the empty space with combs. If the apiarist keeps a swarm too long in a tied bag and is late in hiving it, the bees will start building their first comb right there; the comb will consist of regular worker cells and sometimes the queen will try to lay there while the workers gnaw at the bag to provide an entrance.

A nucleus—a mere handful of bees—builds a little comb of standard-size cells.

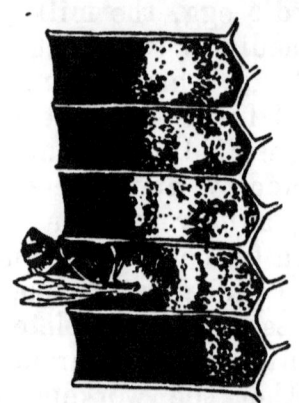

Pollen loads are, as a rule, deposited in cells in which several generations of young bees have been reared, for the remains of cocoons make the bottoms of such cells rounded which enables pollen to be rammed well in a compact mass. The ramming is done with the head, the more compact the mass of beebread, the better it keeps

A living organism's need for definite conditions as a manifestation of its heredity, the heriditary properties of a living organism as its ability to react in a definite way to stimuli, are evident in every detail of the attitude of the workers, the queen, and the drones to the nest, the combs and the cells

Academician T. D. Lysenko says: "A living body builds itself from the conditions of its environment, from its food in the broad sense of the word . . . the conditions of life, the external environmental conditions, on being assimilated, incorporated as component parts of the living body, become internal conditions . . . become conditions essential for the growth and development of this altered body."

At the same time the uniform size of the cells shows the narrowness of the demands the heredity of bees makes on the conditions within the nest. Thus, it is not any kind of cells the bees need to store protein food, or the queen to lay her eggs in, but definitely situated cells of a more or less strictly uniform diameter and shape.

Taken out of her colony and isolated from it, a bee is a far less perfect and much more inferior creature than the same bee in the colony.

Simple experiments show that the nervous make-up of an isolated bee is relatively primitive. A decapitated drone preserves for an appreciable time the ability of reacting by reflex to certain stimuli; a beheaded bee can sting. If the abdomen of a bee sucking nectar from a flower or syrup from a feeder is cut off, she will continue sucking, moving about and lifting her wings.

In the colony, among creatures like herself, that same bee possesses complex and highly perfected instincts, such as the building instinct we have already studied, or the orientation instinct which we shall study later on.

What is the source and the cause of the higher organization of a bee in the hive, in the nest?

It is the colony itself which acts as a factor moulding the nature of the individuals it is made up of.

It is the organization of the colony as a biological entity, itself ascending to a higher state and becoming more complex in the process of evolution, which perfects and enriches each individual with new adaptive traits.

We see from the above that, like all living organisms, plants and animals, bees change the environmental medium of their life as a result of their vital activities. Here we mean the environmental medium of each individual bee, the medium in the primary and narrow sense of the word—the nest. In their turn, the new conditions created in the nest necessarily exercise a reciprocal influence on the authors of those changes.

Consequently, the nest of wax combs in which a bee matures and in the narrow spaces between which she spends all her life, is for a bee the primary condition of the environmental medium, her hereditary necessity, the natural condition of her existence. Themselves products of natural selection, the conditions in the nest mould the bee and train in new bee generations the need for definite conditions.

This need is satisfied and this living condition is recreated with such accuracy and elegance that Darwin had every reason to speak of the "inimitable architectural powers" of the common bee, of the comb being absolutely perfect in economizing labour and wax, of the fact that the bee "had practically anticipated the discoveries of profound mathematicians," and finally, of the cell-making instinct of the hive-bee being "the most wonderful of all known instincts."

We know Darwin's explanation of the perfection with which combs are built, which is that since bees expend several times the amount of energy on wax secretion as on producing honey, the utmost economy in wax is a necessity for the species. That is why the progeny of the colonies which are the better architects, by inheriting this talent, acquire an advantage over others and are better preserved and reproduced. It is but natural, then, that honey-bees, like the Osmia Papaveris, the Ceratina, the Anthophora, and other bee species, may not live to see whether their progeny survive and may not know the importance of the economy in wax they exercise in building the correct combs, and yet the best architectural talents of bees are preserved and in their turn re-create the stereotyped conditions by which they were bred.

Bees, of course, have no more idea that they build their cells at a definite distance from one another than they have of the magnitude of the rhomb angles in the Maraldi Pyramid. But in the course of millenniums natural selection has picked out, favoured, developed, and accumulated those modifications in bee-colonies, which have resulted in the construction of cells of the greatest possible strength and of the proper size and shape with the greatest economy of wax.

Those bee-colonies which made the best cells with least waste of honey in the secretion of wax flourished and transmitted their newly-acquired instincts to new swarms, which in their turn, owing to their economical instinct and the peculiarities of their building habits, grew and developed more successfully than others.

In disputing the helpless attempts of some scientists to explain the building art of bees as the result of the struggle for existence between themselves, K. Timiryazev wrote:

"The building instinct is not a weapon directed against other bees, but only a weapon in the struggle against the conditions of existence. Poor builders, having spent their energies on the secretion of extra precious wax, will store less honey which may not last them through the winter so that they will perish. This will not happen to good builders, who, on the contrary, will spread ever wider, conquering such localities where the winter requires greater stores of honey."

This thesis of winter and wintering deserves a more detailed study.

THE WINTER CLUSTER

The Temperature within the Cluster: the Result of Food of Mass and of Movement. The Flat Cluster in the Observatory Hive. Bees in Amber, and Fossil Bees. The Origin of the Force Drawing Together Biological Species. Conclusion from the Study of the Temperature in a Bee Nest.

Clustering together is a natural characteristic of the honey-bee. Under different conditions and at different stages in the life of the colony, this mutual attraction of bees is determined by different causes.

In hot countries or during the hot season a swarm may make its abode under some kind of shelter in the open air. The outer layer of such a swarm is made up of motionless, apparently inactive bees, clinging to one another. When the outside temperature is about 35°C. this living shell becomes less compact, and at lower temperatures it covers the entire nest, combs and all in a compact mass some three to four centimetres thick. An opening is left in this living shell through which bees come and go.

In winter the mutual attraction of the bees is definitely regulated by temperature.

Almost all insects beat a retreat at the approach of winter, and the most they can do is to put up a passive resistance. They pass the cold spell in a state of anabiosis, not actually dead but only apparently lifeless. A spark of life smoulders in their bodies, kept kindled by fat accumulated during the summer. The warmth of the sun alone recalls the insect to full life, provided the store of fat is sufficient to last it through the winter.

The honey-bee whose fat substance is but poorly developed cannot hibernate. She warms herself in winter by food from without. Here we see an instance of a living organism feeding in order not to die and not reducing its vital functions in order to feed.

It has been long known that animals with smaller bodies lose more heat than bigger ones, the surface of their body being greater in proportion to its size. Indeed, a sparrow or a mouse killed on a frosty day becomes instantly cold while an eagle or a bear remains warm for a considerable time.

Text-books illustrate this geometrical and physical phenomenon by the following computation: the volumes of three cubes with edges of one, two and three centimetres are one, eight and twenty-seven cubic centimetres respectively, while their respective surfaces are six, twenty-four and fifty-four square centimetres. Surfaces of bodies grow in proportion to the square and the volumes in proportion to the cube of the edge. This is why big forms have relatively smaller surfaces than small ones, and at low temperatures small animals lose more heat than big ones.

Depending on food alone, bees could not survive our winters and would be frozen even in a hive with combs full of honey. However much food an individual bee might carbonize within her small body, the heat generated would not be enough to make up for the loss of heat.

But the same laws of physics which doom to death an individual bee stand in good stead to the bee-colony. When in cold weather the bees crowd together in a spherical cluster, the greater the number of bees the smaller the ratio of the surface of the sphere will be to its mass.

The smaller the surface of a body, the less its absolute loss of heat; consequently, the more compact the living shell of the winter cluster, the less its surface, the better the comparatively less compact centre of the cluster will be insulated from the outward cold, the less heat it will radiate. At the same time, the more food the bees in the centre consume, the more heat they will generate.

A special experiment was carried out in order to establish how reliable and effective the heat insulation provided by the outward layer of bees in the cluster is. Several hives were left to winter in the open with metal netting substituted for all but the front walls. Into such a net hive protected from the wind but not from the cold, a colony was

94

put in mid-November. In the locality where the experiment was conducted the temperature was as low as —30° C. in January and —20° C. in February. The experimental colony which wintered practically in the open, without any other protection than its own living shell, lived through the win-

The temperature has dropped. The cluster draws closer together

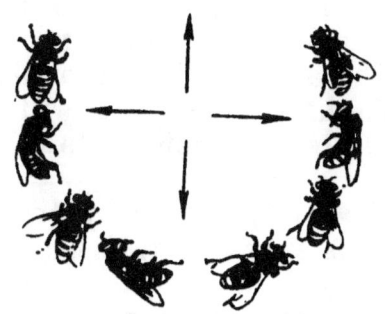

The temperature has risen. The cluster expands

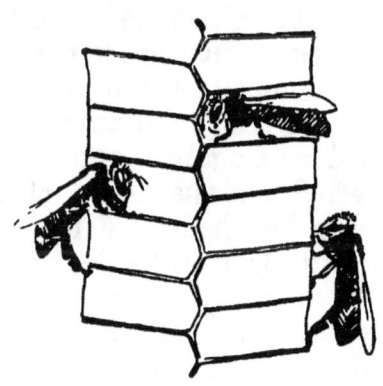

ter. Although during the six months between December and June it lost much of its strength, by the autumn it pulled through and stored enough honey for the next winter.

When the brood-nest is well protected from the cold, wintering in the open does the colony no harm.

The ability of the honey-bee to winter in the cluster alone has influenced her anatomy: the bee has special glands the secretion of which discharged into the rectum retards the decomposition of the waste food matter accumulated there during the winter. This is one of the numerous means bees are provided with for maintaining that cleanliness in the hive for which they are justly famed.

It is the usual practice for bee-keepers in temperate zones to put away beehives in special winter premises where a moderately low temperature is maintained at a constant level. But often bees are left to winter in the open, and if you make your way under the rime-covered leafless trees to the apiary where the hives are surrounded and capped with thick snow, you will hardly believe that under this dead blanket there beats the living pulse of the bee cluster.

By studying, step by step, the winter life of the bee-colony, scientists have established that the living sphere of bees interlayed with comb slowly changes its position, at the same time preserving its shape; it moves along the bee-spaces, unsealing comb after comb and consuming the food in the cells.

The most active bees gather in the centre of the cluster where the queen is. They start vigorous movements as soon as the temperature reaches 14° C.; they eat great quantities of honey which warms them and generates warmth for the others, including the bees on the outer edge, which, pressed one against another, make the living cover of the cluster. Since the bee's chitinous skeleton and the hairs covering her body are bad conductors of heat, this living shell affords a reliable protection for the cluster against loss of heat.

In an observatory hive with only one frame the cluster is, naturally, flat. And here one may see an almost perfect disc consisting of thousands of drowsing bees, immovable on the combs, like the inhabitants of a fairy-tale slumberland. A gentle tap at the hive-wall will awaken the sleepers for a moment, they will flutter their wings in agitation, a slight rustling will be heard in the hive, and in an instant all will be still again.

As the outside temperature falls, the bees consume more food and, consequently, generate more heat in the centre, while those in the outer layer try more energetically to escape the cold and make their way into the centre.

The thousands of ever-moving individuals can by their joint effort raise the temperature within the cluster to

35° C. The centre, where the queen is, can be located in an observatory hive by touch, for the glass is the warmest over it.

Having warmed itself, the colony ceases moving until the temperature in the cluster falls to the critical point of 14° C., when the bees again start warming themselves by feeding and moving.

But stores of highly caloric honey or the mass of the winter cluster taken separately would not be enough to allow the colony to outlive the winter. The bees are able to vanquish cold only by means of active defence — *food, mass, and movement.*

The muscular exertion of a mass of the cold-blooded in order to raise the temperature may be considered a most wonderful "invention" of natural selection.

In their attempts to ascertain the origin of such an adaptive measure as the honey-bee winter cluster, scientists have long been trying to reconstruct the most important stages in the history of the species.

The Hymenoptera genus, to which the honey-bee belongs appeared on the earth about 150 million years ago. It may be noted in passing that the honey-bee is not the most ancient of extant insects — the dragon-fly is much older than the bee.

By reading the stone annals of the world's history from traces left by past eras deep in the earth, paleontologists have been able to learn much that happened on our planet long before the appearance of man. Their findings have thrown some light on the life of the insects coeval with the first flowering plants.

How elementary those insects were compared with our honey-bee! A solitary creature without any permanent nest that collected no food and laid eggs on the first leaf that it came across—that is how scientists describe the remotest ancestor of the honey-bee.

The larvae of this insect, just like many caterpillars of today, ate both the leaves and the flowers of plants, which were pollinated by the wind. The flowers, too, were quite unlike the ones we are familiar with: they had neither nectar in the calyx nor a gay-coloured corolla.

Clouds of pollen were floating in the air and the sticky stigmas of pistils retained the pollen, while the caterpillar-larvæ devoured the nourishing pollen on the pistils. Myriads of enemies preyed on these insects, their eggs and larvae.

Natural selection, which embraces the heredity, variability and viability of organisms, is an ancient law of nature. It began manifesting itself on the earth simultaneously with the appearance of conditions favourable for life, in other words, it was born at the same time as life itself, and since then it has been active, changing, developing and perfecting all living creatures.

Changes in their environment transformed and developed the insects then extant, from whose midst came the Hymenoptera—the forbears of the modern honey-bee.

We have been able to understand and explain certain details in the anatomy and organization of prehistoric bee species only with the assistance of the entomologists who have described the anatomy and behaviour of some modern bee species. Combined efforts on the part of paleozoologists and entomologists have resulted in piecing together into a system the disjointed scraps of the history of the honey-bee.

When flowering plants developed anthers, the tongues of the insects foraging on the flowers became longer, their bodies developed adaptive devices for carrying pollen, no longer the only food of the four-winged, now that there was nectar.

Those were the times of a tempestuous growth of plant life, since it was a hot and humid summer all over the globe. Ever-green oaks, bay-trees, cinnamon trees, magnolias and palms grew side by side with coniferous trees on the Baltic coast, for instance.

Among the traces of such forests preserved in soft sandy rocks of the Tertiary period there have been discovered numerous amber seams—layers of time processed resin of extinct coniferous species. In this amber, specimens of the nature of bygone ages have been preserved intact to our day.

Bees inhabiting the forests long before the Glacial epoch which so materially changed the face of the earth, seem alive in the liquid gold of the mineralized resin. But these nature-made amber preparations seldom contained more than two or three specimens. Then excavations in an old peat bog enriched history with something like four score lime-impregnated fossil bees, on whose bodies even the hairs were preserved.

It has not been established even approximately how many million years ago the bees, which evidently came here for water, were swallowed up by the mud and in what way their chitinous skeletons became impregnated with lime in the bog. Neither has the answer been found to the question why there is a fragment of only one drone and no queen, while all the eighty bees are definitely divided into two groups—big and small bees. It has been established that, though outwardly very much resembling the queen, the big bees have wax plates on the lower side of the abdomen, straight sting and pollen baskets on the hind legs, all of which are characteristics of workers. Probably, those were no longer solitary bees but members of a colony.

Considering that the properties of these bees show certain resemblances of the individuals to our bees, and at the same time suggest a material difference in the organization of their colony as compared with modern bees, we may suppose that the highest stage in the specific evolution of bees is chiefly the perfection of the *colony* as a biological unit.

The finding in the peat we have just discussed throws light on many an important problem in the evolution of bees and brings us to the conclusion that the history of the bee-colony extends over millions of years. During this time the deepest changes both in plant and bee life were wrought by the Glacial period, during which some features of colony life which we have not yet touched upon developed and strengthened.

And now let us consider these features at some length. Some biologists take a one-sided view of the ability of plants and animals to multiply indefinitely, seeing in it only the striving on the part of species to struggle for space under the sun, to spread all over the globe, which they allege, is characteristic of all living nature.

Everyone is well familiar with the calculations often quoted in biological literature concerning the time needed for the progeny of a bacterium, or of a dandelion seed, or of a couple of elephants, to spread all over the earth. But the biologists who were so keen on such calculations often failed to see other forces, those which draw species together.

Why do hungry wolves run in fierce packs in cold winters? A sensible answer given by the hunters is that the wolves attack in a pack a prey that would be too strong for a single wolf.

But why, then, do hares run in herds in cold winters? They do not attack anybody and can in no way help one another.

Why do birds living individually gather in flocks preparing to fly to the south at the approach of winter?

Ornithologists answer that birds migrate in organized flocks because old leaders play an important role in drawing around them young inexperienced birds.

But then why do locusts emerging from holes in barren waste lands make their raids in enormous hosts headed by no old leaders? Why do sea herring gather in immense shoals? The ichthyologists say that this is very natural—fish always spawn in masses.

But then why do fry and tadpoles swim in schools and keep together while they are too young for mating?

Why do May-flies congregate in countless thousands around night bonfires where they are singed and burned in great numbers?

Why does the tortoise-beetle hibernate in thousands under rotten leaves in forest glades, where, to protect the crops from the voracious insect, collective-farm brigade-leaders and insect-fighting specialists drive flocks of fowl which destroy the pest?

The zoologists answer that the places the tortoise-beetle chooses for hibernating are probably particularly well suited for this purpose, as is the warmer and more tranquil water near the bank—for fry, or bright light—for the May-flies.

All these explanations are true, but they do not exhaust all the aspects of the phenomenon.

Take a solitary bee of the genus Halictus. In fine weather the males fly tirelessly about, each by himself, while in bad weather, and also in the evening when dusk gathers they always congregate in groups on a bare branch or stalk. It has been repeatedly stated that they find here no protection from the cold, no food, no females that might prove to be the attraction. Still they gather together for some reason or other, but the conditions fostering this urge of organisms have not yet been studied and clarified.

Might not the struggle between living creatures belonging to various species, might not this alone finally have brought about the development of a centripetal force drawing together the organisms of one and the same species?

As far as bees are concerned, we know something of the way bee-colonies grow and of the factors determining their size, thanks to the researches of G. F. Taranov.

Galvanometers connected with temperature-controlling apparatus placed in various parts of a hive have helped scientists to study the heat-generating properties of the bee-colony and to glean some facts explaining what has made the honey-bee a social insect.

The females of wild solitary bees lay about a score of eggs during their lifetime. The study of the thermal regime of a small colony consisting of merely twenty bees lifts the curtain on the historical—or rather prehistorical—past of the honey-bee. Research has shown that, like solitary insects, this prototype of the primary colony becomes warmer or colder in accordance with the temperature of the air.

The tiny colony was to all intents and purposes at the mercy of the outside temperature, and yet, between it and solitary insects a difference is observable: at 20°C. a handful of twenty bees generates only 1° of its own heat, while at—14°C. it generates 2°. The lowering of the outside temperature makes the bees generate more heat.

The difference is almost insignificant, but one must not fail to see in it an embryo of a capacity to protect oneself against cold.

Like everything in nature, this capacity changed and developed from the simple to the more complex, from the lower to the higher.

At medium temperatures a colony of five hundred bees behaved like any solitary insect, passively gaining or losing heat in accordance with the fluctuation of external temperature. At such temperatures the air within the nest was only one degree warmer than the outside air. But when it became too hot outside, the colony began to resist excessive heat and the nest, which at medium temperatures was

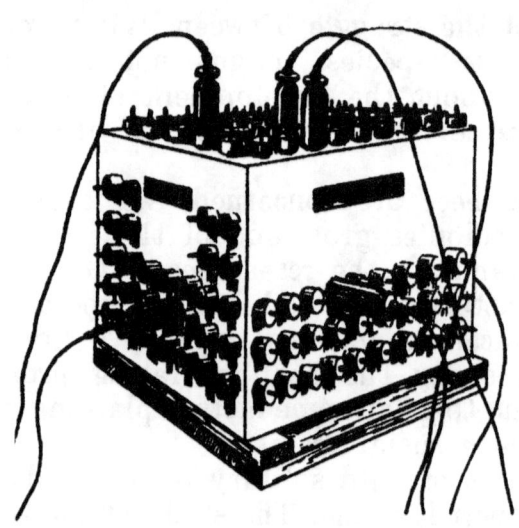

Thermometers have been installed in this hive for observing temperatures in various parts of it

always warmer than the air outside, was unexpectedly cooler than the atmosphere around it. Although the difference was but slight, still it proved that the bee-colony was capable of reducing the temperature in the nest.

Moreover, at temperatures less than 18°C. the little colony of five hundred bees escaped the influence of the weather altogether: while the outside temperature was zero, the cluster was warmed up to 23°C.

This means that at extremely low and extremely high temperatures the condition of a colony of insects differs essentially from that of solitary insects.

This difference is still more profound and more pronounced in a colony of five thousand bees. At extremely low temperatures the temperature in the nest reaches 26°C. and at extremely hot times is reduced by four degrees. In a colony, bees, cold-blooded like all insects when taken individually, have acquired the ability to maintain the nest temperature constant in all weather, thus becoming warm-blooded, as it were.

The colony began producing for itself a very important condition of life—warmth.

What special advantages has the big colony of modern bees in this respect? — was the next question of the researchers.

It has been established with the help of an ordinary precision balance that while warming or cooling the nest, a bee from a small colony expends, on the average, more energy, and consequently, more honey than a bee from a large colony. Calculations have shown a bee from the strongest colony experimented on (35,000 bees) to be six to seven times more "economical" than a bee from a weak one (2,000 bees). It was known before that in weak colonies the bees produce less honey, and now it has become clear how more wasteful such colonies are. The smaller stores of honey gathered by weak colonies are the result of both smaller takes and greater waste. A strong colony is doubly profitable; the bee in it is at the same time more efficient and more economical.

In an experiment, one colony of sixty thousand bees weighing about six kilogrammes stored one and a half times as much honey as four colonies weighing one and a half kilogrammes each and totalling the same number of bees.

Strong colonies show to the best advantage in localities with a poor honey-flow.

It should be borne in mind that, after wintering in cluster and preserving the fertile queen and a sufficient number of workers, the colony enters the spring season in full vigour, ready for hive work and field flights.

Winter is not yet over, but immediately after the solstice bees, even if they have wintered in a cellar where no sun-rays penetrate and where the temperature is maintained at a constant level, will start eating more beebread and feeding the queen with royal jelly. The queen then begins laying eggs in the empty cells. Larvae are hatched from the eggs and the workers rear the first generation of spring bees which are to replace the winter-worn autumn bees.

The higher the sun rises over the horizon the more intensively the queen lays. By the time fine spring weather sets in, young bees emerge in masses from the cells rejuvenating the hive population and swelling the colony, whose compound strength has enabled it to pass through the ordeal of wintering, a task that would have been too much for an individual bee.

A LIVING BRUSH

FLOWERS AND INSECTS

Bee-Hunting and Domestication. What Catalogues of World Flora Showed Us. The Source of the Viability and Conservatism of the Progeny of Cross-Pollinated Parents. Changes in the History of the Vegetable Kingdom. Wind and Insects. Flowers—a Magic Table-Cloth.

From times immemorial and until recently the hunt after wild honey-bees was the only source of honey.

V. K. Arsenyev, the famous explorer, made the following entry in his *Ussuri Territory Diary* in 1906:

"As we were having tea somebody took out a cup in which some honey was left and immediately several bees appeared over the bivouac one after another. Some were arriving while others were hurrying home with their loads to come back for more. Cossack Murzin volunteered to find their nest. He noted the direction in which the bees flew and, holding the cup of honey in his hand, he faced in that direction. In a moment a bee came. As she flew home Murzin followed her with his eyes as far as he could see, then he advanced along the bee's route and waited for the next bee to come; in this way he approached the nest slowly but surely, the bees themselves showing him the way. One must certainly have much patience for such a hunt."

This was written in 1906, but today, too, people take honey from wild bees.

Seeing the traces of a Siberian polecat leading to a hollow tree, the hunter sets a trap on its track. If the polecat caught has paid a visit to a bees' nest this will be seen by its faulty fur, short sparse whiskers and swollen face. If, in addition, dead bees are scattered around the tree, this is a sure sign that there is honey in the hollow. The hunter marks the tree with his mark for all to see that the bees and honey are his property.

In winter, trees harbouring bee nests can be easily recognized by scratches—the traces of bear's claws on the trunk.

In summer, bee-colonies can be discovered not only by means of honey baits as described in V. K. Arsenyev's diary, but by following them from their watering places. While a bee is taking water or foraging, a skilful hunter ties a long woollen thread around her leg. The flight of a bee marked in this way can be easily followed even on an overcast day, on sunny days flying bees can well be detected by the gleam of their wings in sunlit spaces between trees. In addition to the direction of flight the time within which the bee comes back is important, for this enables the observer to calculate the distance from the nest.

Cases are known when in hollows inhabited by bees for a very long time (the age of the combs is determined by their colour, the old ones being quite dark) as much as 300 kilogrammes of honey has been found. But such cases are extremely rare, and usually the take amounts to ten or twenty kilogrammes of hard-gotten honey.

That is why colonies of wild bees discovered in forests are by various means driven from the tree hollows into hives and taken to the bee-garden where one can be sure of obtaining honey with less trouble.

By hollowing living trees in forests to house swarms in man passed from hunting to domesticating this insect. When, however, the first apiarists put swarms into skeps and brought them closer to their homes in order to get more honey and wax, they had no inkling that they were spreading over the areas under crops an insect without which the flowers of many cultivated plants remain unpollinated and do not set seed. But we have cause to think that it was precisely this circumstance that in the final analysis proved of the greatest importance for the development of beekeeping.

When, in our times, economists tried to translate the profit derived from bee-keeping into terms of money, they discovered that while producing a ruble's worth of honey and wax, bees bring about a ten to fifteen rubles worth of crops in the fields, gardens and orchards. So we can justly consider the bee as producing not only honey but also harvests of various crops.

Buckwheat, sunflower, camelina, melon, water-melon, pumpkin, cucumber, apple-tree, pear-tree, cherry-tree, and numerous other cultivated plants bear no fruit if no pollinating insects visit them. In order to form a still better idea of the importance of this factor for plants and all living nature, it would be well once more to glance over the history of living forms.

The catalogues of world flora compiled in the early twentieth century include 176,000 plant species and show that flower plants, of which there are 103,000 species, constitute almost two-thirds of the world's green wealth.

Many agricultural crops produce a harvest only if pollinated by bees

At the same time it is well known that the flower plants are the youngest class of plants. Why, then, does the youngest class contain the greatest number of species?

It has been established that before the "dawn of vegetation" as represented by the simplest green schizophyta, appeared on the earth, half of the known geological periods had passed. The interval from the schizophyta to true plants was traversed much more quickly, but for a very long time the earth was covered with a vegetation tens and hundreds of times more uniform than the present, the few species growing on our planet changing very slowly until the flower plants came into being.

During the Cretaceous period the greatest change in the history of plants occurred: the vegetable kingdom was renovated, the domination of the gymnospermae, of which now only a few hundreds of species remain, came to an end. The angiospermae came to the fore in the evolution of the vegetable kingdom. Their first modest flowers marked the beginning of a new era for the flora.

Bees pollinating flowers are directly connected with this phase of the history of plants, and it is not just a coincidence that paleontologists have discovered and continue to discover prints of insects capable of pollinating plants beside the remains of the first true flower plants.

What was the decisive factor in ensuring the victory of the insect-pollinated species?

"An outrageous riddle" was how Darwin characterized the rapid development of all higher plants during the following geological era. Later, by explaining why sexes and plants of different sexes appeared Darwin was able himself to solve this riddle in the main. Darwin discovered in living nature the great advantage that results from the fusion of two slightly different individuals and established that the progeny of cross-pollinated plants is more vigorous.

This law is of general biological significance and obtains equally in the plant and the animal world.

Soviet agrobiology has made a new contribution towards the elaboration and further development of this branch of science.

Academician T. D. Lysenko has explained how the viability of a new organism is intensified through fertilization and showed that the degree of its viability depends on the difference between the elements united in the process of fertilization, and, finally, that the difference between these elements is determined by the conditions of life of the organisms within which they exist.

It is now clear why cross-pollinated plants yield more and bigger seeds and fruit than self-pollinated plants. It is now equally clear why plants that are the progeny of cross-pollinated parents are hardier and better adapted to the changing environmental conditions.

Mosses, lichen, and ferns whose germ cells are carried only by raindrops, can fertilize female cells with male ones only from near-by plants living under similar conditions. And all such species develop very slowly; thus, modern ferns are known to differ but slightly from those of the Carboniferous period of the Paleozoic era. The only difference is that the modern forms are much smaller.

The flower plants receive pollen from plants living at some distance and, consequently, under somewhat different conditions, and their progeny, naturally, is more viable and possesses greater adaptability. That is why insects, by carrying pollen from one plant to another, were able to accelerate the development of flower plants and make them a class dominating in the earth's vegetation. That is why, as the Darwinists used to put it, "insects transformed the earth into a flowering garden."

We see that with the appearance of insects the flora began to develop at an incomparably quicker pace.

The insect species carrying pollen proved a powerful catalyzer in the development of plant forms and at the same time underwent important changes themselves, as we have seen above.

It is a remarkable fact, and one which, as will be seen further, will help us to understand some intricate problems of bees' history, that a more prolific and more viable progeny of plants can often be obtained even without cross-pollination, just through slight changes in the conditions of their life.

Plant-breeders know full well that the importing of seeds of vegetables and other crops from regions where conditions of growth are different will often exercise a beneficent influence on the variety in question by raising its hardiness and yields. In this instance the different environmental conditions themselves directly assimilated by the organism make it more viable, though not to the same extent as when these conditions are introduced through pollen.

But if such is the case, why are there in the temperate zones so few *wind*-pollinated flower plants? Specialists have computed that the wind pollinates four times fewer plant species than insects, while all of the ancient gymnospermae, the predecessors of modern flower plants, were wind-pollinated.

The reason for this is very simple. The wind is not a reliable pollen-carrier. It carries pollen from one flower to another in a haphazard way and is effective only when there is plenty of pollen, and therefore wind-pollinated plants had to spend an enormous quantity of nutriment on the production of pollen.

Numerous stories can be found in the works of naturalists about extensive areas in regions where wind-pollinated plants grow being covered with a thick carpet of pollen, about pollen carried by the wind high into the mountains and covering snowy fields and glaciers, and about sailors at sea sweeping off pollen carried on board by the wind.

Although nature is unstinting where propagation of species is concerned, still it does away with unnecessary waste, and in this we can see an explanation why insects have become the chief agents in pollinating flower plants.

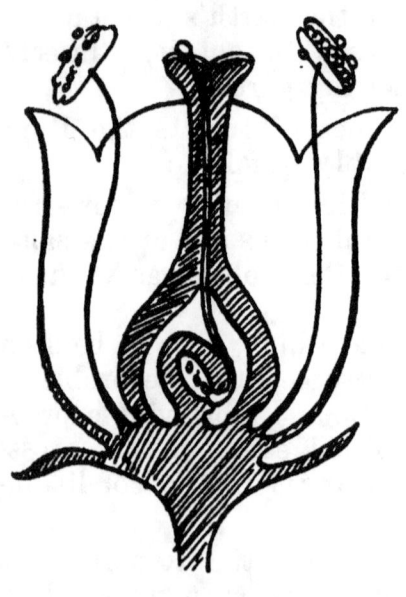

A diagram of a flower showing two stamens (the one to the right with a ripe and burst anther) and a section of the pistil with a pollen-grain on its stigma that has penetrated down to the ovary

Even when insects ate pollen and, flying and creeping from flower to flower chanced to carry grains of pollen on their bodies, as did the remote ancestors of our bees, they proved much more reliable and less expensive pollinators than the wind. The advantages of insect-pollination became still greater after plants started producing nectar and developed flowers with gay-coloured petals and an inviting scent announcing to the sight and smell of the insect the presence of nectar hidden within the flower. It is not just a coincidence that the flowers of wind-pollinated plants are devoid both of scent and coloured petals.

Provided with bright petals and alluring scent, the flowers were more assured of being pollinated, and the insects could find food more easily. Observation of insects on the flowers of plants usually pollinated by the wind has shown that flowers without easily recognizable characteristics are visited irregularly, in a haphazard way.

In his *Historical Method in Biology*, K. A. Timiryazev quotes the example of bees gathering honey from flowers which seem to "work for the bee, preparing food for her." Timiryazev sees in this a visual proof of the fact that "the advantage explained by and resulting directly from natural selection can be either purely individual, egoistic, or mutual. Natural selection offers no explanation of an adaptation harmful for the being it belongs to but useful for some other being."

A plant, naturally, produces nectar in its flower not only to have the flowers of other, neighbouring, plants pollinated by the insects (there are no adaptations useful for some other being alone!) but to attract to itself insects carrying pollen from other flowers.

But between individuals of different species in nature there is not only a struggle but also mutual help. The relations between plants and insects which pollinate them furnish an instance of mutual help.

K. A. Timiryazev says that the pollination of plants by insects explains the mutual advantages of this extremely complex and infinitely varied adaptation uniting the life of plants and that of insects in a harmonious whole.

The various forms of connection between species deserve a closer study.

All animals feed by devouring not only their own progeny but often, which is less known, by devouring themselves.

A wolf eats a she-hare before she has given birth to her young; a hawk carries away a partridge in whose nest lie unhatched eggs; a pike swallows a sleeping carp which has left no progeny; a rabbit nibs at young blades of grass of which no seeds will be left; goats strip mountains bare by eating grass and bushes. With such a system of interspecific relations, an animal feeding is, so to say, cutting the bough it stands upon.

Naturalists suppose with reason that "the predatory economy" of the animals makes them finally adapt themselves to a new kind of food, which invariably results in a modification of all the animals of the given species.

And what about bees?

Bees, which as we have seen, proved to be such a potent factor in accelerating the evolution of the vegetable kingdom, served as a factor in an entirely opposite evolutionary trend.

The more diligent they were in gathering nectar, the more seeds were formed in the flowers they pollinated the more plants grew from those seeds and the greater the number of flowers on those plants, with a corresponding increase of nectar for the coming generations of bees. This may remind one of the magic table-cloth in the fairy-tale, on which the number of dishes increases as you eat.

The feeding habits of bees being fixed, they have a definite stabilizing influence on the sources from which they get honey, and here apparently the cause of the relatively limited variability of bees lies.

Such was the case in wild nature only as long as bees themselves chose the location of their nests in areas rich in various vegetation assuring a stable supply of nectar and pollen.

But everything changed radically when man started burning, felling and stubbing forests, ploughing up meadows and steppe-lands, draining bogs, turning them into fields sown to uniform crops, many of which produce no flowers that might be a source of food for bees.

On the other hand, man began to concentrate in his apiaries tens and hundreds of bee-colonies for which big areas of nectareous plants were needed. The natural steppe and forest steppe, a rich foraging area for bees, were turned into *field-steppe* sown mainly with cereal crops which yield nothing for bees. What is more, the number of apiaries in these field-steppe areas is very great.

Under the mode of production prevalent in the past in Russia and today in capitalist countries only the immediate, direct results of human activity were taken into consideration. This applies both to social and natural consequences of man's actions. The more remote and later consequences, however, as the classics of Marxism noted long ago, often proved entirely different, quite unexpected and often counteracted the former.

This led to the number of bees which needed food increasing quickly while the foraging area decreased still more quickly. As under this double influence the natural ties between the insect-pollinated plants and the pollinators were being severed, the magic table-cloth lost its miraculous property. Thus even the best of bee-keepers were faced with the problem: where the bees were to gather honey, and even: what the bees were to be fed with?

Neither the sowing of honey-plants for bees to forage nor the migration of apiaries in search of honey-bearing plants growing in various localities could eliminate the contradiction that had arisen. It became possible to solve the problem effectively and to re-establish the severed connections only under planned socialist economy, and gradually at that. We shall speak of this further on.

FOREIGN POLLEN

Advantages and Disadvantages of Bisexual Plants. The Black Blood: a "Trisexual" Plant, and the Sexual Variations of the Avocado. Grain Crops in the Steppe and the Coffee-Bush on the Island of Guadeloupe. Artificial Fertilization of Plants in the U.S.S.R.

Wherefore glowest thou crimson, flower?
Wherefore sparklest in dewy flame?

These are the words the poet A. Koltsov addressed to a flower in one of his poems. In his time few were acquainted with the works of A. T. Bolotov, an outstanding Russian astronomer and naturalist, who, in the middle of the eighteenth century, proved to be far ahead of the scholars of all other countries in understanding the material essence of fertilization in plants. In his early work *Essay on Apple-Tree Seeds*, Bolotov wrote: "At the blooming time the blossoms of the apple-tree are daily visited by a multitude of bees which fly from tree to tree in search of honey. To make 'beebread,' the bees collect loads of yellow pollen which one may see very well on their hind legs. It may easily happen that, by brushing with their pollen-covered legs against the unfertilized pistils in flowers which have not yet been fertilized by their own pollen, the bees enable nature to engender. . . seeds in the flowers."

Almost two hundred years ago, A. T. Bolotov, an outstanding Russian naturalist, explained the biological significance of bees' visits to the apple-blossoms

In his other works, particularly the articles published in *Economic Magazine* in 1780, Bolotov passed from hypotheses to a definite statement that "for all its flowers and germs, a plant sometimes cannot produce fruit and seeds, at least the latter, perfect enough for the reproduction and propagation of the species, unless there occur certain necessary events not always depending on the actions of the plant itself. Often these depend on entirely foreign factors, such as other plants, air, wind, dew, and sometimes, insects." "Seeds in fruit are engendered," Bolotov wrote in another article, "not only with the help of the wind but also through the agency of some insects, and especially bees crawling over flowers in order to gather their nectar . . . and beebread. Bees gather pollen from many flowers in the pollen-baskets on their hind legs, but moving in a flower they drop it on the pistils and thus convey it where it is needed."

Christian Conrad Sprengel, Fritz Müller and, which is not generally known, Joseph Gottlieb Koelreuter made a valuable contribution to the study of the relation between flowers and bees by publishing in the *Acts of the Russian Academy of Sciences* the results of his observations and the experiments performed in St. Petersburg Botanical Gardens, which showed that insects take part in pollinating plants, that nectar is the means of attracting insects, and that bees make honey from nectar.

Darwin continued the study of this problem in his experiment with Linaria, one of the first attempts to determine the biological consequences of insect pollination. In this simple experiment, Linaria seedlings obtained through self-pollination and through cross-pollination were grown on two beds, one beside the other.

"To my surprise," Darwin wrote, "the crossed plants when fully grown were plainly taller and more vigorous than the self-fertilized ones. Bees incessantly visit the flowers of this Linaria and carry pollen from one to the other; and if insects are excluded, the flowers produce extremely few seeds; so that the wild plants from which my seedlings were raised must have been intercrossed during all previous generations. It seemed therefore quite incredible that the difference between the two beds of seedlings could have been due to a single act of self-fertilization; and I attributed the result to the self-fertilized seeds not having been well ripened, improbable as it was that all should have been in this state, or to some other accidental and inexplicable cause. During the next year, I raised for the same purpose as before two large beds close together of self-fertilized and crossed seedlings from the carnation, *Dianthus caryophyllus*. This plant, like the Linaria, is almost sterile if insects are excluded; and we may draw the same inference as before, namely, that the parent-plants must have been intercrossed during every or almost every previous generation. Nevertheless, the self-fertilized seedlings were plainly inferior in height and vigour to the crossed." This started the long series of well-known experiments, which Darwin carried on for more than ten years, and which showed that an overwhelming majority of plants require cross-pollination and suffer from self-pollination.

A by-product of Darwin's researches was the discovery of innumerable and highly varied ways and means by which the plant world protects itself from harmful self-pollination and provides its flowers with pollen from other plants.

There are species, which like goat-willow and hemp, are bisexual and dioecian: some specimens produce only male flowers, others only female. With such plants cross-pollination only is possible. Then there are monoecian plants with flowers of different sexes, as, for instance, maize, pumpkin, and melon. But it is in neither the plants' nor the bees' interest to have flowers of different sexes and it is not very prevalent in nature. The reason is that bisexual flowers are visited by insects collecting both pollen and nectar while unisexual plants and flowers attract only half as many insects.

With bisexual flowers, each visit of a bee may result in pollinating it

Even if an insect visits the male and female flowers of a species indiscriminately one after another, only 50% of the visits are useful while with bisexual plants each visit of the insect may result in pollinating the flower. From this it will be seen that at least twice the number of insects is needed for pollinating unisexual flowers. That is why species with bisexual flowers have become so wide-spread.

At the same time, thousands of devices have been developed to prevent self-pollination in such flowers. The pistil in the flowers of the lime-tree, for instance, ripens after the stamens have ceased to produce pollen; the stigmas in the flowers of lupine and alfalfa are covered with a pellicle; the pistil in the hazel flower ripens before the stamens. At the risk of remaining unfertilized the flowers of the crimson clover do not accept either their own pollen or that of other flowers on the same plant but wait for insects to bring pollen from other plants.

The flowers of some fruit-trees can be pollinated by pollen not merely from other trees, but from trees of some *other variety*. The flowers of certain bisexual trees open not simultaneously, but one after another, beginning from below, so that when ripe pollen falls from the upper flowers the lower ones have been pollinated and are proof against self-pollination.

The oak-tree has unisexual flowers, but the female flowers grow in the upper part of the tree and the male flowers, in the lower, which rules out self-pollination.

There is no end to the diversity of the extremely "clever" anatomical and physiological adaptations of plants protecting them from self-pollination and ensuring the supply of foreign pollen to the flowers.

Among all such devices the most interesting for us are the various details of mutual adaptability and correspondence existing in the structure of a flower and the body of the insect that pollinates it. The exquisite anatomic and, as it is becoming increasingly evident, physiological mutual adaptability of flowers and insect-pollinators leads us to believe that insects are, to some extent, derivative of flowers, and this should not be understood to mean merely that the body of an insect is made up of the substance derived from the plant it feeds upon.

In their turn, plants, too, have more or less adapted themselves to the insects without which they cannot reproduce.

This connection suggests the conclusion that there exists a common link in the formation of the heredity of flower plants and of their pollinators, a vitally-important factor, present in the development of the two parties to the pollinating process.

What can this mutually-important factor be?

Bees gather from flowers only pollen and nectar, which at-

Borrago officinalis is a major honey-plant of the central regions of the U.S.S.R. Bee-keepers say that bees lick nectar off its flowers as from plates

tracts the insects to the flowers but is not required for the fertilization of the plants. At the same time pollen, without which, as a rule, seeds do not set, is indispensable to bees as food for the brood and for the nurse-bees feeding the queen. It is but natural, therefore, to suppose that the development of the mutual adaptability could proceed through pollen. Consequently, it may be possible to adapt plants and insects mutually by artificial means in those cases when man wants their mutual connection to be more full and complete.

Now for a close study of the mutual adaptability of insects and plants.

It has long been known that after visiting a flower like sage or orchid a bee carries pollen which the stamen, working like a lever, has attached to her body exactly in the spot from which it is sure to pass to the pistil of the flower the bee visits next.

It is also well known that buckwheat produces bisexual flowers of two kinds—one with short stamens and a long pistil and the other with long stamens and a short pistil. Cross-pollination between the two different forms results in perfect seeds, truly corresponds to cross-pollination. In crossing the flowers of the same type the seeds are scarcely better than in enforced pollination. The cowslip with its two types of bisexual flowers, one more like the male and the other more like the female, the crossing of which produces perfect plants, presents a similar example.

Black-blood is a more complex case. Those who have wandered in damp meadows are familiar with its crimson clusters with as many as three types of flowers: long-tubular, medium-tubular, and short-tubular. This is something like a tri-sexual plant, in which each form has flowers with pistils and stamens, but one form is like a female, the other, a male, and the third, intermediate between the two. The differences in the length of the pistils and stamens in each type make possible six pairs of combinations in cross-pollination producing viable progeny.

Fewer people are familiar with the avocado, whose buttery fruit has earned it the nickname of "cow-tree."

It is being acclimatized in the Soviet Union on the Black-Sea coast, south of Sochi. All avocado plants are divided into two groups, outwardly absolutely identical, which are distinguished one from the other only by the behaviour (mark: behaviour, and not form) of their bisexual flowers growing in clusters. The flowers of Group A, as horticul-

The pictures given here and innumerable similar facts bear witness to the premise that while being derivative from plants they pollinate, insects, in their turn, are a prerequisite of the existence of plants

turists call it, accept pollen only in the morning when they produce no pollen, those of Group B, only in the afternoon. So a tree belonging to Group A is female in the morning and male in the afternoon and Group B the opposite way about. Successful cross-pollination is possible only between trees of the different groups. This type of pollination was discovered but recently and offers one more example of nature's inventiveness in trying to avoid self-pollination which, in Darwin's expressive words, she abhors.

There are many more cases where the structure and physiology of flowers of bisexual plants are highly conducive to cross-pollination. The flower elects beneficent foreign pollen which infuses vigour and viability into the progeny.

All the same, there are plants capable of fruiting through self-pollination.

Darwin explained why nature created and continues to create self-pollinating plants. In order that the propagation of the species might be effected, it is often better for a plant to be pollinated by its own pollen than to remain unfertilized if no foreign pollen is brought either by insects or by the wind. Biologists have long been aware of this.

Self-pollination has certain advantages as a means of preventing sterility.

In one of his works, Academician V. L. Komarov says that "in many localities where insects are scarce and winds abundant, as for instance in the steppe, flower plants have had to re-adapt themselves to wind-pollination and simplify the structure of the flowers."

Sometimes even in tropical and sub-tropical countries where insects are plentiful, plants suffer from lack of pollinators. The coffee-bush, for instance, is insect-pollinated in a wild state. But when large plantations of this bush were planted on the Island of Guadeloupe, nature could not provide a sufficient number of insects and it became a typical wind-pollinated plant.

We know that buckwheat flowers are dimorphous, the two forms being equally rich in nectar

But why speak of the Island of Guadeloupe, lying so far away from us? We know of facts close at hand showing how urgent the need of insect-pollinators is becoming for our developing agriculture with its numerous field crops.

Bees willingly visit the sunflower from which they gather rich crops of pollen and nectar, but the collective farms growing sunflower have to resort on ever-increasing areas to artificial pollination in addition to pollination by bees.

While the sunflower is in bloom, women collective farmers walk between the rows and with soft mittens made of rabbit-skin brush the golden flowers, carrying pollen in the fluff from one flower to another. Thanks to this, more and bigger seeds are set, and larger crops gathered in.

Additional artificial pollination of buckwheat is also widely practised.

Wherever this method is used in rearing crops the yields are bigger, which is convincing proof of the necessity of increasing the number of insect-pollinators.

It is clear that when the collective farms have more and larger apiaries, there will be no need of additional artificial pollination for a number of crops. This will be all the more useful as insect-pollination, particularly bee-pollination, has one more advantage of which we shall speak in the next chapter.

Succory is an important honey-plant. It is an industrial crop grown for the sake of its roots from which substitute coffee is produced and which are used as raw materials for the production of sugar and alcohol

A MIXTURE OF POLLEN

Wheat and Rye in Bloom. "Marriage for Love" in Plant Life. Bees—the Most Important Pollinators of Non-Cereal Crops. Selective Visiting of Flowers. The Advantages of Repeated Natural Pollination with a Mixture of Pollen.

Several years ago Academician T. K. Kvaratskhelia started observation of an apple-tree grown on the plot of Mukhrani University, near Tbilisi.

When the flower-buds became swollen in spring, the tree crown was divided into three parts with an equal number of buds in each.

One third was enclosed in a textile sack preventing insects from visiting the flowers within, which all were pollinated artificially.

Another third was covered with a large gauze net and a beehive was put in the enclosed space, so that the bees could work only the flowers within.

The remaining third was left open to insects and the wind.

When the blossoming was over the ovaries were counted with the following results: there were set ovaries in one-sixth of the artificially-pollinated flowers, in one-third of those in the open part of the tree and in one half of those under the gauze, where the bees had worked.

By the autumn there were very few fruits on the boughs of the artificially-pollinated part of the tree and the branches were directed upwards; the branches of the open part were slightly bent down. But in the part where bees had been at work the branches hung low to the ground weighted by apples.

At the time this experiment was carried out, it was not clear why out of a hundred flowers scrupulously pollinated by hand only sixteen set fruit, while out of a hundred identical flowers pollinated in a natural way—by bees—the fruit-setting occurred in nearly sixty flowers.

Today research in this field has made great progress, the problem being particularly well worked out in the case of cereals.

Wheat is a self-pollinator and before the flower has opened, the ripe anthers of its thin stamens leave pollen on the early-ripening fluffy stigma; the pollen then germinates in the direction of the ovary. The male and female cells unite and wheat seed is formed.

So well does wheat pollinate itself with its own pollen that in crossing two different varieties of wheat plant-breeders must remove unripe stamens from flowers that are not yet completely developed. This timely surgical interference prevents self-pollination and the castrated flower on which foreign pollen is deposited with a fine brush, can produce hybrid seed, the fruit of the two varieties chosen for crossing.

It would seem that the wheat flower avoids foreign pollen and accepts it only for want of its own. Often the progeny of this "forced marriage" is inferior, less viable and weaker than either of the parents.

How many pages in the history of wheat selection are devoted to descriptions of the disappointments of hybridizers who, after crossing two excellent varieties, received worthless hybrids.

But perhaps the example of wheat is a refutation of Darwin's premise that long self-pollination is harmful? No, it is not. Wheat is but a further illustration and confirmation of the rule.

Darwin himself knew that after self-pollination wheat flowers open and exsert anthers.

"Mr. Wilson believes that all the pollen shed by the exserted anthers is absolutely useless," Darwin wrote in answer to one of his correspondents. "This is a conclusion, which it would require very rigid proof to make me to admit."

Michurinists supplied this rigid proof. The result of it was that Darwin was right in doubting Mr. Wilson's statement.

Of what use can this belated production of pollen be to wheat? Pollen has germinated on the stigmas of all the pistils and the seed is setting, why, then, does flower after flower exsert its anther?

This is quite unlike rye. At the quietest hour of the morning tiny anthers burst open soundlessly and here and there over the motionless rye-field rise compact cloudlets of pollen. More and more little clouds appear, expand, and merge together; the rye-field undulates and seems to breathe in the dry hazy cloud of pollen sticking to the stigmas of the pistils. We know that rye is a cross-pollinator, but why does wheat, a self-pollinator, produce as much pollen as rye?

It has been proved that it is not in vain that after self-pollination has taken place, wheat pollen is carried by the wind. This pollen is used for the additional pollination of the recently self-pollinated flowers in the ears.

Thus we see that in wheat self-insurance against sterility comes first and defence against its own pollen, second. Rye, on the contrary, puts protection from its own pollen first and comes to fruition only if pollinated with foreign pollen. Rye flowers artificially fertilized with their own pollen either remain sterile or produce very few seeds. Such seeds germinate badly and the plants from them are weak and non-viable.

When, however, a lone rye plant was grown among wheat where it could have used only its own pollen, it produced excellent seeds and the plants from the seeds were strong, quite unlike the weaklings usually received in the result of self-pollination. It looked as if the admixture of wheat pollen had paralysed the harmful action its own pollen would otherwise have produced.

Experiments have brought to light one more interesting detail. As long as rye plants received additional wheat pollen in a natural way they prospered, but when isolated rye plants received pollen from that same wheat by means of the hybridizer's brush the seeds in the ears were meagre and the plants from them stunted, often just like the cripples obtained through usual enforced fertilization of rye by its own pollen.

The stunted and weak rye plants with small ears offered an illustration of I. V. Michurin's words that if a plant is ". . . freely choosing—out of all the wind- and insect-carried pollen, sometimes from quite a few different varieties— the pollen that is best suited to the structure of its particular fruit-forming organs, the progeny consists of relatively more viable individuals. The latter cannot always be expected in hybrid seedlings, which are obtained by artificial and, of course, enforced crossing."

It is astonishing how unexpected and varied are manifestations of the harmful effect produced in plants through enforced fertilization by unsuitable pollen.

Academician A. A. Avakyan crossed winter wheat Hostianum-0237 with the spring varieties Erythrospermum-1160 and Erythrospermum-1163 (the latter two wheats being full sisters.) When flowers in the ears of the winter Hostianum were fertilized by the pollen of the spring wheats, the effect of the two pollinators was identical. The Hostianum ears produced seeds which ripened and germinated fairly well: the first green leaf was followed by a second, then a third....

And then a mysterious thing happened: as soon as the third leaf appeared, some incomprehensible fault of the plant organism made the first shrivel up; the fourth leaf appeared and the second shrivelled up. The same was observed with the fifth and the third, and so on. This happened not to an individual plant in the hybrid progeny, but to every one of them. All of them dragged out a miserable existence with two or three leaves and eventually perished.

Then A. A. Avakyan fertilized a Hostianum-0237 plant with the mixture of pollen taken from spring wheat Erythrospermum-1163 and a winter wheat—Ukrainka. The seeds obtained produced a normal progeny similar in many respects to Ukrainka.

In another experiment, Hostianum-0237 was pollinated first with the unsuitable Erythrospermum-1160 pollen and subsequently with the pollen of Lutescens-0329. The hybrid progeny was normal and resembled Lutescens.

In one more case a plant of that same Hostianum was pollinated by the pollen of Erythrospermum-1160, but this time it was obtained from plants of this spring variety which for two generations before crossing had been sown not in spring but in autumn. The result of crossing with plants grown in different conditions proved different; the offspring of the crossing grew and developed quite normally.

Such is the significance of elective pollination. When it was discovered, Michurinist selectionists jokingly spoke of "marriage for love" in plant life.

The research of Soviet agrobiologists, particularly Academician D. A. Dolgushin with wheat and Academician I. S. Varuntsyan with cotton plant, has contributed to a better understanding of how electiveness in pollination is manifested by plants.

This research has given rise to a new and important branch of the science of heredity.

Standing at the cradle of the variety, as it were, it was possible to observe the development of its likes and dislikes, to see what determines them and causes their changes. Many important facts for this new branch of genetics were supplied by experiments in artificial crossing of plants.

Hundreds of such experiments, not only with self-pollinated wheat, or with cotton plant, but also with wind-pollinated maize, insect-pollinated sunflower, buckwheat and many grain crops, industrial crops, fruit-trees, decorative and other plants, have unanimously shown that, as a rule, viable and healthy progeny is obtained from a plant only when it can freely choose pollen for fertilization. This is a prime requirement both in hybridization and in pollination generally.

Satisfactory results were obtained in experiments with different plant varieties only when sufficent quantity of pollen grains was introduced into the flowers. The fewer the number of pollen grains the worse the seeds and the plants obtained from them.

In an experiment with water-melon, not one flower that received from three to twenty-five grains of pollen set seed. The only one flower that received twenty-seven grains bore fruit, but it was under-sized, misshapen and contained almost no germinable seeds.

In an extensive plot badly visited by bees eighty per cent of the water-melon ovaries died off, and the fruits that developed were very small, weighing not more than one kilogramme and containing small seeds. The water-melons were covered with spots suggesting some unknown disease.

Melons, pumpkins and cucumbers in that plot also suffered from lack of bees. Almost all ovaries on the plants died off and the fruits were misshapen out of all recognition:

the melons, which were no bigger than a pear, were crooked, the cucumbers of the celebrated Nezhin variety were also crooked and wrinkled.

But when an apiary was moved close to the plot, all the newly-formed water-melons were big—up to five kilogrammes each—and free from spots. The newly-set melons, pumpkins and cucumbers, too, were of normal size and shape.

Patient observers counted that a female water-melon flower in the plot near the apiary was visited by an average of thirty-six bees, each of which had visited at least twenty male flowers before. Consequently, bees brought to one female flower pollen from hundreds (the exact average figure is 720) of male flowers. Although other species of the melon family require a smaller number of pollen sources, still the pistil of a pumpkin flower receives pollen from about fifty male flowers, the pistil of a cucumber flower, pollen from two hundred male flowers, and the pistil of a melon flower, from five hundred male flowers.

Pollination with a sufficient quantity of pollen invariably yielded good fruit while pollination with a limited number of pollen grains brought about results similar to those obtained through the usual enforced pollination of flowers with their own pollen.

The same was evident in experiments with cotton plant.

K. K. Yenikeyev, a well-known horticulturist and plant-breeder and a pupil of Michurin, carried out experiments with stone-fruit trees—cherry, plum and apricot.

In studying the causes of the low crops from Vladimirskaya (Roditeleva) cherry, K. K. Yenikeyev established that, after pollination with a mixture of pollen from two varieties the yield was three times bigger than with self-pollination. A long series of experiments proved that each time Vladimirskaya cherry was pollinated with a mixture of pollen from two varieties such as, for instance, Plodorodnaya and Lyubskaya, or Shubinka and Sklyanka Rozovaya, or Krasa Severa and Polevka, the yield was always bigger than when pollen from only one variety was used.

The conclusions K. K. Yenikeyev arrived at in the

course of his experiments with cherry were confirmed by experiments with dozens of other species: pollination with a mixture of pollen results in a better crop and raises fertility.

The effect of pollinating with a mixture of pollen is not limited to directly affecting the yields obtained from pollinated flowers.

In an experiment at the Michurin Institute on three neighbouring beds stones from varieties of the Skorospelka Krasnaya plum obtained through pollination with the pollen of Reinclaude Kolkhozny, Reinclaude Reforma and a mixture of pollen from the two Reinclaudes were sown at the same time. On three other beds seeds of Skorospelka Krasnaya obtained through pollination by Reinclaude Kolkhozny, Tern Krupnoplodny and a mixture of the two were sown.

A great number of different variations were thus sown and tested at the institute. The annual measurements of the seedlings showed that in almost all cases pollination with a mixture of pollen produced more viable seeds, more vigorous shoots, and healthier, taller and more uniform seedlings.

In some cases two-year-old seedlings differed by as much as thirty centimetres, the seedlings from mixture-pollinated seeds being taller by a third than their brothers on the neighbouring beds.

All these facts should be well noted.

The wider the scope of experiments, the greater the number of different plants studied by agronomists and plant-breeders, the plainer it became that a mixture of pollen possesses greater fertilizing potentiality and that an abundance of pollen from various sources works miracles with plants.

But why is free natural pollination often more effective than artificial pollination, even when the latter is done with a mixture of pollen?

Can it be by mere coincidence that with many plants the percentage of seeds set is much higher in the presence of insects?

In one of his letters Darwin mentions Erythrina, an Australian plant, which produces no seed unless in pollinating the horticulturist repeats the movements of the petals caused by bees.

In experiments with buckwheat, alfalfa, and sunflower it was proved that artificial fertilization was more effective if the stigmas were slightly scratched with a needle, the scratching probably substituting the rubbing of the insect's chitinous body against the stigmas of the flower.

These facts afford one more proof that bees do not merely carry pollen from one flower to another.

A question may arise: why do we speak here of the honey-bee alone? Is she the only insect that pollinates the flowers of plants? Do the hundreds of thousands of other species play no part in pollination?

Flowers are indeed visited by a great variety of insect species which specialists divide into three groups: chance visitors, conditional visitors and obligatory visitors.

The main pollinators are the insects belonging to group three which, in its turn, is subdivided into pollinators of wide, medium and narrow spheres of action. Bees occupy first place

Willow-herb or fireweed. It would be hard to find another honey-plant in the northern and central parts of the U.S.S.R. and in Siberian taiga equal to willow-herb in the abundance of nectar and the long period of blossoming

among the obligatory visitors with a wide sphere of action.

It is worth while to choose a tree on a sunny spring day when apples, pears, cherries and plums are in bloom, and observe what insects come to forage on the fragrant blossoms.

Although bumble-bees and butterflies pollinate flowers, they are by far inferior to honey-bees as pollinators of agricultural crops. Here are (from top to bottom): garden bumble-bee (a), stone bumble-bee (b), and the Machaon butterfly (c)

Such observations are very instructive also in a vegetable garden, if we turn our attention to plants that are grown for their fruit as, for instance, cucumbers or vegetables from which a crop of seeds is to be obtained, and not such as are grown for their leaves (cress, cabbage, tobacco) or roots (carrot, beet, radish).

The meadows at the blossoming time of the grasses—food for domestic animals—present the same picture as the orchard and the vegetable garden: we see that bees are the chief pollinators of the majority of agricultural crops.

It has been calculated that bees constitute eighty per cent of all visitors to the flowers of cultivated plants, only twenty per cent falling to other insects—various wasps, flies, bumble-bees, beetles, butterflies and moths. These twenty per cent include a great number of insects which, in visiting plants, cause harm to them. Even in cases when these visitors feed on pollen they effect no pollination, for pollen does not stick to their smooth chitinous tests.

Almost all butterflies and most flies are at some stage of their development destructive insects, and always bad pollinators of flowers to which they are attracted only by nectar.

True, some butterflies flying from flower to flower may pollinate them, but they stay so long on each flower that they can visit very few plants during a day. Numerous flies, and also such insects as Cynipidae, Cephidae, and Ichneumonidae are of little use as pollinators.

As for wasps, bumble-bees and various species of bees, in gathering pollen and nectar they visit flowers of one species and even variety for several days, they work the flowers much quicker than do butterflies, and their fuzzy bodies become covered with plenty of pollen grains.

Domesticated honey-bees possess the greatest abilities and possibilities as pollinators. Their flying season begins in early spring and ends in autumn and under favourable conditions they visit flowers all that time, while wild bees do so only for a few days in the season.

The honey-bee lives in colonies numbering tens of thousands of insects, while wasp or bumble-bee colonies seldom exceed two or three hundred. In spring, when wasp and bumble-bee females start their colonies, bee-colonies, having survived the winter, begin their pollinating activities at once. Fruit trees, as a rule, bloom before the first worker bumble-bees emerge.

It is true that bumble-bees have the advantage over honey-bees in that they can forage on cold and windy days. But, honey-bees differ from bumble-bees in being able to visit flowers deprived of their petals by wind or rain. Besides, and this is very important, honey-bees which man has put in a hive, can be transferred by cart, lorry, raft, boat or plane to any place where plants need pollinating.

And this is not the least of the advantages of bees as pollinators. The majority of insects visit flowers to satisfy their own hunger and provide food for their young, and however voracious, an insect cannot eat more than it can consume. The honey-bee can carry enormous quantities of nectar to the hive in her small honey-sack. In hives where the bee-keeper regularly removes honey-filled combs and replaces them by empty ones, the colonies work incessantly and become tireless visitors and pollinators of plants.

All these advantages are the more essential because, as has been stated above, bees are not mere mechanical carriers of pollen from one flower to another.

Everyone who has had a chance of observing insects on flowers has seen that upon leaving the flower from which she had been sucking nectar, a bee would make a jerky and rapid flight to the one next it. She seemed to fall on the flower of her choice but immediately rebounded without touching it, as though thrown back by some invisible force; then again approached it as if unable to decide, and

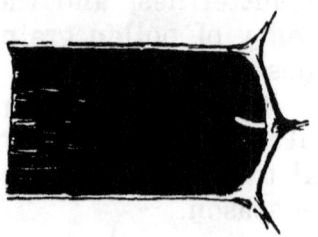 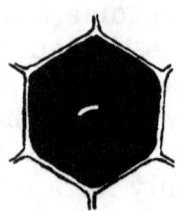

The egg just deposited by the queen lies on the bottom of the cell like a semi-transparent white fleck of dust, or a fragment of a snow-flake. Even people with good eyesight sometimes fail to discern the eggs in the cells

finally left the flower altogether. Many people have seen this but few have pondered over the causes of the bee's whims.

Why, indeed, did the bee refuse to visit the flower? A simple examination established that the flowers of the clover plants which the bee had rejected were affected with mould, although outwardly they seemed quite sound. Some time later symptoms of the disease became apparent to man.

The bee's circumspection in refusing to visit a flower which was discovered to be diseased furnishes fresh proof of the acuteness of her instinct and shows that without caring for the flower, of course, the bee is discriminating in pollination and visits only sound plants.

The honey-bee's peculiar qualities as a pollinator are manifest not only in her electiveness in visiting flowers. The two pollen loads a bee carries on her hind legs to the hive may weigh as much as twenty to twenty-five milli-grammes and contain from three to four million pollen grains. To collect them, the bee must visit several hundred flowers—about three hundred in the case of clover—and carry on her back from flower to flower a large quantity of grains. In the heat of a summer day as many as 50,000 to

75,000 pollen grains were counted on the hairy body of a bee. And if we bear in mind that each flower is visited not by one but by several bees, we can easily imagine what a great number of most varied pollen grains is brought to the flower.

Moreover, one and the same flower is visited by bees several times at different stages of its development and at different hours of the day. This influences the properties of both the stigmas and pollen and cannot but influence the development of the plant. To this must be added the influence of the admixture of pollen from other varieties and species which the bee has visited casually. There are undoubtedly many more factors, remaining secrets to this day and awaiting their discovery.

All of the factors, both known and unknown, taken together, provide an answer to the outrageous riddle that puzzled Darwin who saw the benefits of cross-pollination but did not know many of its important aspects. Today, thanks to the research of Soviet agrobiologists, they are being revealed and to the biologist the bee is like a living brush, with the help of which flowers are naturally pollinated. Mastering and perfecting this operation is necessary for man who must carry out Michurin's behest and create plants better than Nature.

A TURN OF THE SPIRAL

METAMORPHOSIS

An Egg Laid in a Cell. Three Days Later. Larva Grows
Minute by Minute. "Controlling" the Time of Larval
Development. The Food and Vitality of the Bee-Colony.
Six Days after the Hatching of the Larva. Pre-pupal
and Pupal Stages. Birth of the Bee.

A bee-colony is something more than just the bees we
see on the alighting-board at the entrance and in the air
in front of the hive, going to the field and coming home
with their loads, or moving to and fro along the bee-spaces.
Besides the part of the colony that we see there live, grow
and develop as many, and sometimes twice as many, bees
whose existence is not suspected by one who looks into
the hive for the first time in his life.

About five thousand cells in the combs of a strong
colony are occupied in spring by eggs, something like ten
thousand by larvae, and about twenty thousand by pupae,
so that nearly 35,000 cells are filled with bees "in various stages of unfinished production," as economists would
put it.

Pearly-white eggs, 1.5 millimetre long and weighing
less than 0.1 milligramme, have been attached by the queen
to the cell bases with an adhesive substance. During the
first day after laying, the egg stands at right angles to
the midrib of the comb and parallel to the cell walls; on
the second day it slightly inclines, and on the third day it
lies prone on the base of the cell. During the three days a
larva has matured in the egg, and as soon as the tiny white
annulated worm of a larva hatches, it instantly begins consuming the royal jelly that the nurse-bees have been pour-

ing into its cell while it was in the egg. During the six days of its life, the larva covered with a soft chitinous skin and weighing less than a milligramme consumes 200 milligrammes of food.

While the larva is fed royal jelly (during the first two or three days) it is quite white, but on the third or fourth day a thin vein becomes visible through its transparent skin. Sometimes the vein is of a pronounced yellow. This is due chiefly to the remains of undigested pollen which the larva is fed beginning with the third day. The remains of food are accumulated in the larva's body, since its digestive tract is not yet connected with the rectum. This detail, which ensures cleanliness in the cell, is very important for the colony: if, besides feeding larvae, the nurse-bees had to "wash their diapers," brood-rearing would take up too much time. It should be borne in mind that, from the first to the last moment of its existence, a larva does nothing but consume and digest food which nurse-bees put into the cell day and night.

During the first twenty-four hours after hatching from the egg, the larva increases in weight about five times; in forty-eight hours it weighs nearly 30 times, and by the

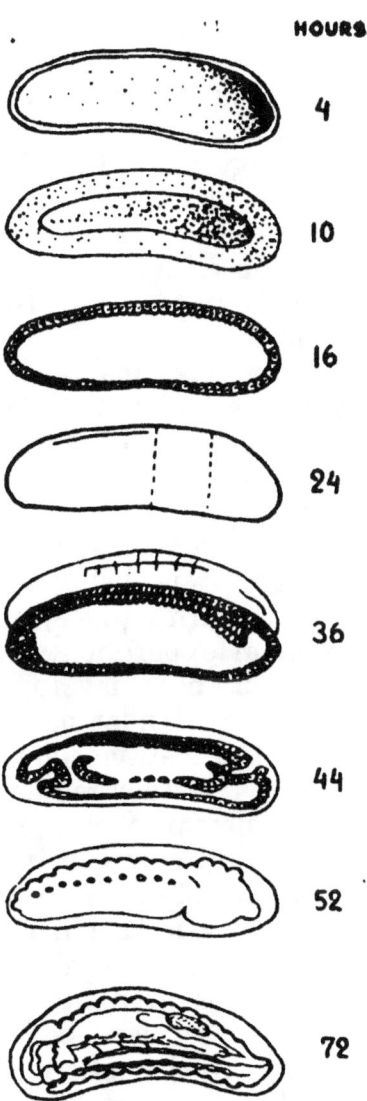

HOURS

4
10
16
24
36
44
52
72

The integument covering the one and a half millimetre long egg that the queen has laid in the cell, is almost transparent. By the end of the third day (to be exact, 72 hours after laying) a tiny white larva matures in the egg

135

end of the fifth day, i.e., of the larval stage, nearly 1,500 times as much as at the beginning. At the end of the pupal stage a queen larva weighs almost 3,000 times as much as upon hatching from the egg! Biologists rightly call the larval stage the "digestion stage." And, indeed, only continuous consumption of food and lack of movement make possible such a rapid growth as that observed in the case of honey-bee larvae. We can safely say that bee-larvae grow not from day to day or from hour to hour, but from minute to minute.

In an experiment at an apiary of the All-Union Academy of Agricultural Sciences in Gorki Leninskiye, a three-day-old larva was taken out of the cell and artificially fed with the help of a pipette under room temperature for seventeen days. It thus lived three normal terms of larval life but did not grow and outwardly remained a three-day-old larva. This was owing to lack of the necessary conditions, and in the first place, warmth.

Special experiments were carried out to determine the periods of bee development under temperatures lower than normal. Under natural conditions with a temperature of 35°C. the metamorphosis took 21 days; at 30°C. it lasted 25 days. A further reduction of temperature by 5° resulted in the quiescence of the majority of the pupae while the few bees that emerged in spite of the low temperature were wingless cripples. At 20° no development of the brood occurred and all the pupae perished.

When, on the contrary, the temperature in the brood chamber was artificially raised, the development periods of the young bees became considerably shorter. At 37° the pupae in the cells matured one or two days earlier than usual, but the bees had underdeveloped wings, were sickly and crippled.

This is why in a normal colony a constant temperature of about 35°C. is maintained in the part of the hive occupied by brood. In the peripheral parts of the brood chamber where the mass of bees covering the combs is less dense and, consequently, produces less warmth, the larvae hatch from the eggs somewhat later than at the end of the third day and their development lasts more than six days.

Exact observations on 2,000 cells in a comb where the queen laid eggs within 24 hours showed that some bees developing in those cells began emerging on the nineteenth day; about a thousand bees emerged on the twentieth and twenty-first days, 500 emerged on the twenty-second day, and the rest continued to emerge for three subsequent days.

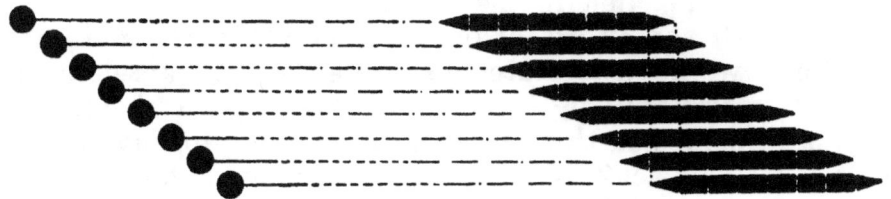

Each circle in this diagram represents a day's laying of eggs (the number of eggs laid each day is, of course, different). The unbroken line from the circles represents the egg stage, the broken line, the larval stage, and the dot-and-dash line, the pupal stage. Mature bees hatch from eggs laid on the same day not all at once but during several days. A. S. Skorikov, a famous Russian expert in bee biology, the author of this diagram, shows with its assistance how from eggs laid on different days young bees may emerge on one day

The above facts clearly show that the processes of growth and development of organisms in living nature are not identical, that the calendar and biological ages of a living individual can be relatively different—depending on the prevailing conditions.

This law is widely applied in plant-growing for accelerating the development of plant organisms by the pre-sowing vernalization of seeds.

Bee biology offers an example of the utilization of this law for the opposite effect. In queenless colonies the bees intensively warm the areas containing emergency queen cells, at the same time warming outlying areas less intensively. One may say that the bees hold up for a time the development of the eggs and young larvae in these areas. We think it timely to mention these facts here, for but recently some scientists observed certain differences in worker-bees belonging to one and the same colony, and declared them to be inexplicable.

A numerous progeny is the result of a single act of fertilization of one female by one male; this progeny has been reared from eggs laid at the same time under equal conditions, the eggs became larvae receiving equal food in equal quantities and developing at a constant temperature, humidity and light. Such being the case, how could differences have appeared in the progeny?

"For all the efforts of our scientific thinking and imagination, we have to state the existence of a causeless and arbitrary variability," an entomologist said helplessly. His argument was this: given a numerous progeny, the result of a single act of fertilization of one female by one male, a progeny reared from eggs laid at the same time under uniform conditions, which later became larvae and were fed uniform food, developed at a constant temperature, humidity and light, how, then, could differences have appeared in it? Besides, no one ever succeeded in creating uniform conditions for an artificially-conducted experiment as those existing in nature.

Granted that this is so, but the queen laying without interruption, eggs laid on different days develop, naturally, under somewhat different conditions, just as was the case in "one-day seedlings" of T. D. Lysenko's Ganja experiments.

By sowing small quantities of seeds of one variety on neighbouring strips every day, Lysenko placed each day's crops under slightly different conditions. The plants on the strips, forming a kind of a living calendar, behaved in different ways, the difference being at times so marked as to make them quite unlike one another.

The "one-day seedlings" are to this day used in plant-growing as an unsurpassed means of biological analysis and of revealing the heredity of a species.

Now let us see what takes place in the beehive where the queen lays every day and where the brood development is always more or less uniform under the conditions of constant temperature and humidity. Here a direct influence of environmental conditions changing with time is excluded. This influence is indirect, reflected, and manifests itself in the condition of the colony in response to external influences. In the final analysis, here, too, each hatch of bees emerging on the same day is a kind of "one-day seedling."

The maintenance of constant temperature and humidity in the hive is a most important element of brood-rearing. Fanning bees expend many times as much energy as bees at rest, but on a hot day when the sun overheats the hive, thousands of bees cool the hive by fanning, and on a cold day thousands of others become nurses: clustering together on the combs they warm the brood and themselves by rapid movement of their legs.

Every hatched larva is an object of constant care. Some bees just look into the cells and pass on, others pop their heads into the cells as if the better to see what is going on there, others again almost completely disappear in the cells and feed the larvae there, unseen by the observer. The older the larva the more often is it fed. A very dense crowd of bees surrounds the queen cells: no sooner does one nurse-bee leave the cell than another takes her place.

The larva consumes the food continually brought by bees into the cell, and grows, becoming more and more rounded in shape. The honey-bee larva moults every thirty-six hours, shedding its skin when it has grown out of it.

We have not yet been able to observe in detail how worker larvae are fed, but we know that nurse-bees visit a larva nearly 1,300 times a day, about 10,000 times during the six days of the larval stage. Queen cells are visited by nurse-bees still oftener.

It would be pertinent to explain here why these details are so important. The food the larvae receive is, as we see, *a mixture of foods*, and this mixture acts in a way similar to pollen mixture in plants. Let this not be taken as a mere play on words: direct experiments have shown that worker-bees developed from larvae fed by small nuclei are far shorter-lived than those fed by strong colonies with a large population and a great number of nurse-bees.

The importance of the number of nurse-bees on the breeding qualities of queens has been proved. The agronomist M. Z. Krasnopeyev selected sixteen colonies of equal strength at a collective farm in Muchkap District, Tambov Region, and bred in them various numbers of queens, from three to fifty. It is clear that the more queens a colony reared, the smaller was the number of nurse-bees attending each queen. As was subsequently proved, the best of the queens reared under this experiment were those fed by the greatest number of nurse-bees. They laid twice as many eggs as those reared by the smaller number of nurse-bees and headed colonies that produced more honey.

The explanation of this phenomenon is not that in the colonies where many queens were bred they were underfed. We must not forget that on the bottom of the cell from which a mature queen emerges there is always some unconsumed food left. This being the case, it seems that the same quantity of food given a larva by a different number of nurse-bees is qualitatively different. M.Z. Krasnopeyev's experiments furnish one more proof of the advantages of strong colonies and reveal one more factor by means of which a sufficient number of bees in a colony was maintained through natural selection. In these experiments each queen was still fed by hundreds of nurse-bees, but if it were possible to reduce the number of nurse-bees still more, the action of the food mixture in strong colonies would be still more evident.

The biology of swarming, too, provides proof of the importance of food mixture, but of this we shall speak later. For the present, we think it worth while to consider another aspect of the biological action of food mixture.

We have seen above how plant species strive to avoid as it were self-pollination and to what variety of devices they resort in this striving. The biology of all forms in the animal kingdom, too, offers well-known examples of nature's abhorrence of incest.

As Darwin wrote "*close* interbreeding diminishes vigour and fertility . . ." and "no organic being fertilizes itself for a perpetuity of generations. . . ."

Today no one doubts the correctness of Darwin's conclusions: with close interbreeding *any* variety, *whatever its characteristics and peculiarities*, becomes less vigorous and sterile. With bees, however, it seems that nothing prevents a drone and the queen he impregnates to come from the eggs laid by one and the same queen and be reared and fed by one and the same colony. Such close interbreeding is unavoidable under many conditions. What characteristics and peculiarities enable bees to escape the evil consequences of such crossings and to preserve vigour and fertility?

Before speaking of them, let us recollect that the flowers of red clover do not set seed if they are pollinated by their own pollen or by the pollen of other flowers of the same plant. In an experiment by Michurinists, a young plant of red clover was divided into two in spring, each half being grown under different conditions: the soils in the pots were different and different fertilizers were added to them. When the plants began to blossom, the experimenters pollinated part of the blossoms with pollen from other blossoms of the same half-plant and another part with pollen from the other half reared under different conditions. The result was that the blossoms pollinated with the pollen from the same half-plant were sterile while those pollinated with the pollen of the other half set seed.

We have mentioned that stunted and non-viable plants were obtained from rye through self-pollination. But Michurinists divided a young rye plant into several parts and grew each in different conditions. When the parts of the plant eared and blossomed they were mutually pollinated. Strictly speaking, this was self-pollination, since all the apparently different plants were closer than brothers, closer than twins—they were parts of the selfsame plant. But the pollination yielded viable seeds from which normal, perfectly sound plants were obtained.

Thus we see that rearing under different conditions and feeding different food helped to eliminate the harmful consequences of close breeding.

This is exactly what happens in the bee-colony: from the third day of their larval life and throughout the pupal and mature life, the queen and the drone are fed differently. The queen receives almost exclusively royal jelly while the drone ceases to receive it from the fourth day of his larval life, feeding on honey and beebread to the end of his days.

Bees provide selectionist plant- and animal-breeders with proof of close breeding being conceived by nature as a means of preserving heredity. At the same time bee life shows that the method of close breeding by parents reared on different food and in different conditions has been tested and found reliable by nature. Even when the parents are closely related by blood such a difference in their rearing will ensure the progeny a certain degree of viability.

Classics in biology, however, have repeatedly warned practical selectionists against excessive use of close breeding. Bee biology teaches a good lesson in this respect, too, for it proves the necessity of caution and circumspection in crossing breeds related by blood.

How can we explain the inevitable death of the drone that has impregnated the queen? His death may be considered a biological adaptation of the bee-colony to prevent the possibility of the drone's repeated participation in reproduction.

In the given case nature seems to detest most the crossing between parents and their direct offspring. The possibility of such a combination in bee life is ruled out under natural conditions. But crossing within the next degree of relationship—the offspring of one parent pair—is quite possible. True, the closeness between two individuals born of the same parents is very great and in other animals the crossing of such couples brings about harmful consequences very soon. Bee biology, however, provides a number of adaptations which either make the realization of such a possibility not unavoidable or eliminate its bad effects.

First, it should be noted that when a queen makes her mating flight, drones from other colonies join the crowd pursuing her and she may mate with a strange drone.

Secondly, the harmful effect of possible close breeding is mitigated by the mixed food the larvae receive.

Now let us come back to the worker or drone larva, which quickly becomes bigger and weighs more than any of the nurse-bees attending it. A mature larva weighs from 150 to 300 milligrammes and completely fills the cell, which becomes too small for it. Only the queen larva has enough space in its big queen cell.

During the five or six days of incessant food absorption, the larvae accumulate enough energy for the coming metamorphoses.

On the sixth day the larva stops eating and straightens out. Little spikes holding the larva in the cell and preventing it from falling out can well be seen if enlarged.

The larva lies in its cell with its head towards the opening. Its stomach, which has been filled with food all this time, at last becomes connected with the rectum, and a tiny drop of undigested pollen is stuck to the cell base. Then the larva starts spinning a cocoon, covering the cell walls with "bee-silk" and the worker-bees seal this "pupa-house" with a porous wax cap.

Bee-keepers wanted to observe the larva spinning its cocoon but for that it was necessary to see what goes on under the cap of porous wax. Sure enough, they could have cut out from the comb several hundred larvae at various stages, as is frequently done, and observed them. But the helpless creatures cannot bear being touched and would have perished. The experimenters wanted to see a living larva spinning its cocoon and not just dead larvae killed at different stages of development. So they contrived to have larvae develop in crystal cells.

Transparent cells, just like wax cells, were kept in a thermostat at the necessary temperature. Here the observers could see every move and turn of the larva, having passed through which it becomes quiescent.

A queen larva does not cover itself completely with the cocoon: the last segments of the abdomen remain free. This peculiarity of the queen cocoon is connected with the presence of food on the cell base and is of major importance for the life of the whole colony.

It takes a queen larva one day to spin its cocoon, a worker two days and a drone three.

In general, drones develop more slowly and pass through all the stages at a slower rate.

In their houses of silk and wax, sleeping pupae spend two days (queen), three days (worker), and four to five days (drone). During this time all that was the larva becomes dissolved into a liquid mass by ferments; all its organs are destroyed and liquefied in the process of transformation. The living matter they were made of continues living, and out of it the cells, tissues and organs of the future bee are built anew from antennae to sting, from head to legs.

Nothing is left of the larva in a mature pupa. The pupal stage is a stage of outward immobility and of radi-

On the third day after hatching from the egg the larva lies circled and completely filling the basal end of the cell. On the fourth day the larva begins to straighten itself preparing to pupate

cal inner changes. The pupating larva weighed about 162 milligrammes and a three-day-old pupa weighs only 141 milligrammes. The pupa has been perfectly motionless and yet it has lost 21 milligrammes, so entomologists seem to be right in saying that the pupa is a "stage of consuming reserves." Under the milky-white chitinous skin, impenetrable to the eye, the plastics accumulated by the larva undergo a transformation, and finally the amorphous prepupa becomes at last like a real, though immovable, bee. The white worm that a few days ago went to sleep in the cell and during all these days has received not a drop of food, will soon awaken in a perfectly new form.

Freed from its skin, the true pupa—the future bee— is colourless and almost wingless. It has considerably lost in weight and is acquiring the last properties and characteristics of a bee. Pigmentation and wings will appear at the final stage before emergence. The compound eyes will appear on the head as purple spots and become darker, the thorax, and the abdomen, too, will assume pigmentation.

Like a wax-ripe seed, the imago's body is soft but gradually it becomes more elastic. The wings are last to appear under the pupal skin. It is amazing how neatly the young bee is packed in its cell. In each and every cell the position of the imago is the same: the head is slightly bent and presses the unfolded long proboscis against the thorax with the ends of the antennae forming a triangle

The larva becomes a pupa which in its time will become a bee. The mature bee gnaws the capping of its cell from within

under the proboscis. The first pair of legs is folded on the thorax and seems to hold the end of the proboscis.

At last the moment comes when the adult insect wakens and sheds the pupal skin. In a queen cell this usually happens sixteen days after the laying of the egg, in a worker cell twenty-one days, and in a drone cell, twenty-four days.

The bee's busy life begins at the moment of emergence. She gnaws her way into the bustling hive through the gossamer cocoon and the wax capping.

However often a bee-keeper may have seen young bees emerging on the combs removed from a hive, he always observes this event as if it were for the first time.

He witnesses the sudden opening of a seemingly lifeless sealed cell; queen cell and drone cell caps open like the round shutter of a port-hole and the worker cell cap is gnawed through and punctured, and from the opening there emerge a pair of cautious feelers on the mobile head of a big-eyed creature covered all over with grey fuzz.

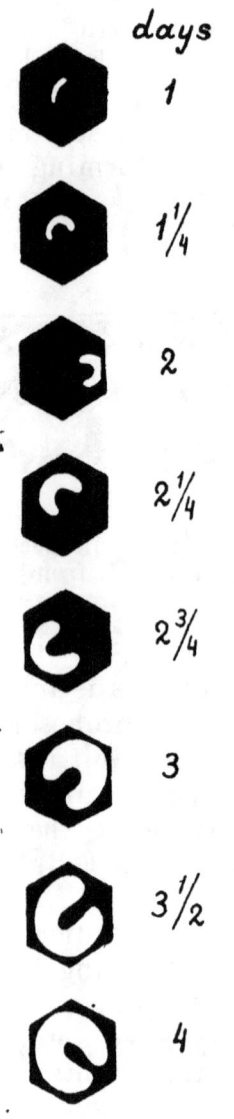

days

1

1¼

2

2¼

2¾

3

3½

4

One glance at the size and position of the larva is enough for an experienced bee-keeper to tell its age pretty accurately

The creature makes many clumsy efforts to get out of its hexagonal cradle and seems to relax now and then to collect strength for a new effort. At last, resting its fore legs on the edge of the cell, it gets out and starts walking about, stretching and grooming itself with all its six legs. It folds and unfolds its wings, waggles its abdomen and seems to look at itself from all sides, the young bee's neck being so flexible as to enable her to turn her head almost by 360 degrees.

While one bee is crawling out of her cell, the neighbouring cell opens and through a crack in its cap an antenna of another young bee appears moving in the air before she herself emerges in her turn.

With a queen laying 1,500 eggs a day, a new bee must emerge in the colony every minute.

Strictly speaking, the emergence of a bee is not her birth. Who gave birth to the new bee? The queen that laid the egg? The builder-bees that collectively built the waxen womb in which the egg ripened? The nurse-bees that day and night gave food to the hatched larva and warmed it like brood-hens? Or was it thousands of workers that from morning till night carried nectar and pollen into the hive to provide food for the brood, for the new generation?

Besides, every new-born being becomes separated from its mother sooner or later, while in the colony of many thousands a new bee becomes part of the mother-community that has reared it.

In addition to the well-known cycle of changes common to all insects—egg, larva, pre-pupa, pupa, adult (or, as entomologists say "perfect") insect — a young bee is to undergo in the community one more cycle of development, one more cycle of metamorphoses: the sequence of duties.

This cycle begins for an adult worker at the moment she leaves the cell and lasts for several days until she makes her first flight to the field where flower-buds are bursting open preparing sweet lures for nectar-carriers and awaiting pollination.

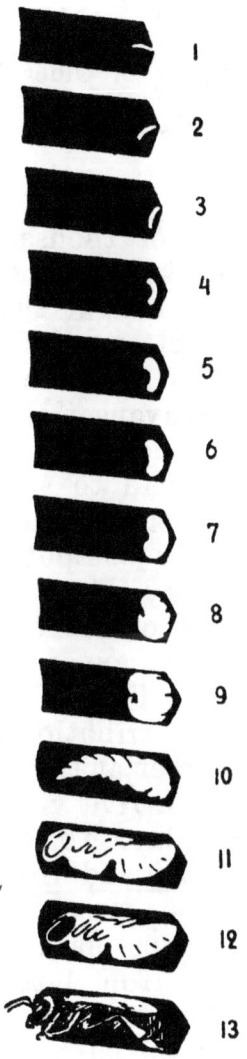

The 21-day-long development of a bee from egg to mature insect is shown here in its nodal stages. 1-3 egg; 4-9 larva; 10 pre-pupa; 11-12 pupa; 13 perfect insect

CYCLE OF DUTIES

A Hive Inhabited by Marked Bees. Cell-Cleaning Bees. Nurse-Bees of Older Larvae. Builder-Bees. Nurse-Bees of Younger Larvae. Home-Bees. Guard-Bees and Scavenger-Bees. Change in Behaviour and Change in Organization.

The tens of thousands of workers in the colony are very much like one another. For a very long time, however, it was thought that all the bees in the colony were divided into classes, guilds and castes and that upon leaving her cell a bee became a forager, a guardian, a nurse, a tanner, a scavenger or a builder for life. To find out what was correct and what was wrong in this common belief, bee-students had to spend many days at an observatory hive into which a group of newly-born bees bearing varicoloured marks was put every day. From their first days in the observatory hive marked bees revealed hitherto unknown secrets of bee communal life. Such was the case in the well-known experiments of Rösch and in the experiments at the Tula Experimental Apiary which made a noteworthy contribution to bee-science. L. I. Perepelova, a scientific worker at the apiary, marked thousands of newly-born bees with five various colours and put them into an observatory hive. To mark the bees, Perepelova and her colleagues used and improved old five-mark code. Colour spots on various parts of the bee's back were taken for various figures. Thus markings on the upper left part represented the figures from 1 to 5, on the upper right part from 5 to 10. A white spot on the left was 1 and on the right 6; a red spot on the left was 2 and on the right 7; a violet spot was 3 and 8; an orange spot 4 and 9; a green spot 5 or zero. The tens were marked on the lower part of the back.

On seeing in the hive a bee with a red spot on the lower left-hand part of her back and a violet on the upper right-hand part L. I. Perepelova knew it to be No. 28; a bee with an orange dot on the lower left-hand part and a green on the upper right-hand part was No. 40. When necessary, bees could be marked beyond the hundred (the fore part of the abdomen) and even beyond the thousand (the lower part of the abdomen). Files were kept for all the marked bees, and observations were entered hourly.

L. I. Perepelova put into the observatory hives 50 "scout" groups of from fifty to five hundred marked bees each.

Bee-keepers say: "The bee-keeper must know his bees as if he has lived in the hive himself," and the remarkable thing about the Tula experiments was that they literally led the observers into the hive where each of the thousand "scout" bees could tell them all they wanted to know.

While observing the work activities of the marked bees L. I. Perepelova kept watch on individual cells in the combs.

A net was traced with a diamond on the glass wall of the hive, the contours of the net coinciding with the pattern of the comb. This net was divided into zones and formed a plan or chart, enabling the observers

Extricating all three pairs of legs from the narrow cell is the most difficult thing. This done, a young bee gets out without much effort

both to study the behaviour of individual bees in the hive and to see whether the work of the home-bees is connected with particular cells or spots on the comb.

This is what the researchers established in the result of their painstaking efforts.

It came to light, in the first place, that while outwardly very much alike, bees in the colony are markedly distinguished by their activities. As we have already said, bee-keepers had suspected this long ago, but they believed that each bee was born for certain duties and spent all her life performing it. Now it became clear that the bee's duties change with her age, and life in the bee hive appeared in a new light.

Let us observe for as long as possible a single bee marked with luminescent paint at her birth so that we can see her on the combs even at night.

In addition to the old and perfectly practicable spot-code, bee-students use other methods of marking bees. A tinfoil circle bearing a number is attached to the back of a bee. Any mark is good for marking groups of bees

On emerging from her cell the mature bee first brushes off the remains of her pupal skin and two or three minutes later begins cleaning the cell. She helps her elder sisters to gnaw off bits of cappings, smooths the edges of the cells, and cleans the inside of the cells with her tongue, for if vacant cells are not prepared for new eggs, if they are not thoroughly cleaned inside, the queen will not lay eggs in them.

The time needed for cleaning a cell so as to make it acceptable to the queen is between 21 and 62 minutes and it has been established with the help of a chronometer that intervals in the work, too, are very uneven, probably depending on the working qualities of each individual bee. In any case, an hour or two after the emergence of a young bee the vacated cell is licked and polished clean ready for a new egg.

It takes from fifteen to thirty bees to clean each cell and prepare it for occupation; the ages of the bees under observation differed widely ranging from one day to three weeks.

L. I. Perepelova observed that while the older bees worked for two minutes at a stretch the young bees rested after a few seconds' work. Often they crawled into empty cells and stayed there for some time, either completing in this way their development or resting from their first work. Many young bees spent some time on brood cells.

On the fourth day of her life the bee ceases cleaning cells. So far she has received food from other bees, now she begins feeding both herself and older larvae, thus becoming a nurse-bee.

She makes her first visit to the combs in which honey and beebread are stored and spends several minutes there, after which traces of her jaws are seen on the smooth surface of the beebread in the cells.

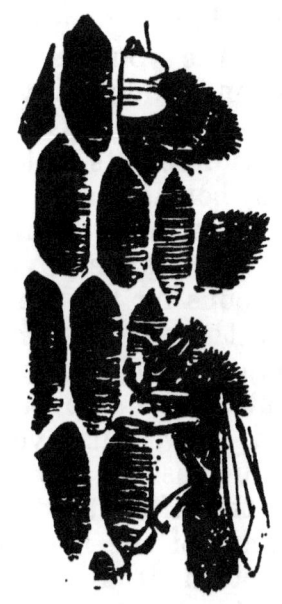

The older a larva the oftener is it visited by nurse-bees feeding it with a mixture of honey and beebread

Taking a load of food the bee hurries to the brood-comb; there she looks into one cell after another and stays for some time in those containing older larvae. Her supply of food spent, the nurse-bee again goes to the stores to get more honey and beebread for a new round on the brood-combs.

The bee performs these duties between the fourth and the eighth days of her life, after which she enters upon a new stage, becoming nurse for the younger larvae. Feeding younger larvae—and they are fed only royal jelly—is the main duty of bees 8 to 12 days old.

L. I. Perepelova has not seen in normal conditions a bee under 6 days of age perform the duties of nurse-bee for younger larvae; the reason for this is now perfectly clear; anatomists, histologists and physiologists have proved that the submaxillary food glands secreting royal jelly are well developed only in bees of a certain age. In a bee less than six days old the glands are still developing, and in one thirteen days old they begin to atrophy; accordingly, nurse-bees feed the younger larvae and the queen when they are between 8 and 12 days of age, when their food glands function best of all. Thus it has been established that a direct connection exists between the physiology of the individual bee, her age, and the functions she performs in the colony.

A similar connection was discovered in the activity of wax glands. Over 500 builder-bees were taken from the festoons on the combs, and since all the bees were marked and the age of each was known, it was established that the majority were from 12 to 18 days old. Sixty thousand preparations of the bees' wax glands were made and studied and it was discovered that the wax glands begin visibly to increase on the twelfth day of a bee's life, stop growing on the fifteenth day and then gradually atrophy. The wax glands of the majority of twenty-three-day-old bees do not function at all.

All these facts once more proved that the bee-colony is organized on a physiological basis and showed how changes in the physiological make-up of the bee determined by her age call forth changes of her duties in the colony.

We have seen our young bee, together with older bees, cleaning cells and nursing brood, first supplying the older larvae with honey and beebread and then feeding the younger larvae with the secretion of her brood-food glands. Then she enters on a new stage of her development.

Upon finishing her duties as nurse-bee the home-bee begins to receive honey and pollen brought by field-bees. The duties of home-bees are varied: some meet the foragers at the entrance and take from them the nectar they have brought, other carry the nectar deeper into the hive from the low-lying cells near the entrance where it was temporarily stored. Still others ram with their heads the pollen brought into the hive by the pollen-gatherers while these are taking honey before starting for a new load.

A home-bee is engaged in storing nectar and pollen for about a week, after which she gradually goes over to house-cleaning duties. In removing all kind of debris from the hive the "charwomen" of the beehive do not just drag the objects beyond the entrance but fly some ten or twenty yards away to drop them there.

As was recently established, bees of this age perform one more duty for the community: for a day or two they act as "barbers," day and night passing from one bee to another and grooming them with the brushes on their legs, smoothing each hair on the head and back with their man-

If you see a bee with her pollen-baskets filled pass her load of nectar to another bee, not far from the entrance, it is a fair guess that this is a forager and the other, a home-bee

dibles. Sometimes the toilet of one bee takes as much as five minutes. From time to time the "barber" interrupts her work, cleans her own mandibles, and again looks for a bee in need of cleaning.

This search for a "client" is in itself very amusing: the "barber" runs swiftly over the comb and strokes the bees she meets with her feelers until one of them responds with a quivering of all her body.

It often happens that two "barbers" clean one field-bee simultaneously.

One cannot help smiling at seeing the groomed bee lift and spread her wings to enable the "barber" to brush the nether side of the wings, the spots between thorax and abdomen, and the back, in a word, all parts of her body which she herself cannot reach either with her mandibles or with her legs. And it is exactly to these spots that some of the bees' direst enemies, such as lice, attach themselves.

Some of the movements of the "barber" in this operation remind us of those made by a pollen-gatherer. The bees clean other bees just as pollen-gatherers clean themselves.

We may suppose that this work prepares for new duties a bee that has ceased to receive loads from fielders but still works as a scavenger.

The last duty of the bee at home is guarding the hive. Guard duty is performed by bees of all ages with the exception of very young ones, but the best and most vigilant guards are the older home-bees and old foragers.

Guards at the entrance are formidable-looking creatures

As has been said, guard duty ends the bee's work inside the hive. After this she enters on a new stage of development, her longest and last— she becomes a field-bee.

If we follow the changes occurring in an adult bee in time and space, we shall see at once that her instinct and the physiology of development imperatively urge her into the bustle of hive life and carry her farther and farther from the centre. The rate at which these changes take place depends on various circumstances.

The scheme of a bee's life may be likened to a bacterium or fungus colony which spreads from the centre, carrying from it cells which grow older as the distance from the centre increases and die on the colony's border.

After hatching in the central zone of the combs, the bee performs her first duties there, and as she grows older the radius of her activities increases and the course of events carries her farther from the centre. First she works in the zone where honey and beebread are stored, then at the entrance, on the border line with the outer world, and finally she crosses the threshold and flies abroad, where she meets her end.

We have just described the cycle of the bee's social duties, as it proceeds under normal conditions. We shall now deal with the recently discovered way of changing this course artificially. The discovery was made by research workers studying problems which seemed in no way connected with the one in question.

Experiments were carried out to study the effect produced on bees by various narcotics such as ether and chloroform, which were formerly used to put queens to sleep for artificial insemination. Carbon dioxide was proved the best for this purpose, since its influence on the insects was instantaneous, rather strong, and they recovered from its effect easily. And while the methods of applying the narcotic were being tried on worker-bees it came to light quite unexpectedly that young bees treated with carbon dioxide skipped the duties appropriate to their age, became older as it were and started collecting nectar. The influence of the treatment was still deeper in the case of queens.

A barber-bee will assume any position to reach with her mandibles some hidden part of the other bee's anatomy

After two treatments with the dioxide (at an interval of 24 hours) the young virgins developed prematurely and began laying without a mating flight, their eggs being of course unfertilized. Usually an unmated queen starts laying unfertilized eggs at the age of not less than a month.

It seemed as if the short 20 minutes spent in the carbon dioxide had made the insects several weeks older.

This reminds one of green fruits, the ripening of which is accelerated by ethylene treatment, but in the case of bees this acceleration was less patently manifest.

A bee awakening after carbon dioxide treatment spends a long time, sometimes a whole hour cleaning herself. She uses all her three pairs of legs one after another; she cleans her eyes and antennae with especial care. Although no outward changes are noticed in the bee, dissections show that her inner organs undergo deep changes after the treatment: the natural course of the development of the glands is disturbed, and a young bee becomes physiologically old.

It should be borne in mind, however, that all the above data based on L. I. Perepelova's observations are generalized, average facts, and it would be wrong to think that the life-cycle of each individual bee proceeds according to a strict time schedule even under normal conditions. Of course, this is not an iron law but a pliable system.

The location of the most important glands in the body of the bee. The glands are never at the height of their development at one and the same time and the changing duties a bee performs in the colony depend on the condition of her glands. 1. brood-food gland; 2. pharyngeal gland; 3. post cerebral gland; 4. thoracic gland; 5. wax-glands; 6. scent-gland; 7. poison-gland

The life stories of individual bees in the observation hive were, naturally, different, but still the above-described laws were observed in the behaviour of bees of the same age.

These laws were discovered in the following manner: it was observed that in all registered cases of feeding older larvae, were engaged bees not less than three days and not more than thirteen days of age. Among the nurses, bees between four and eight days old made up 84 per cent, bees less than four days old made up 6 per cent, and more than eight days old 10 per cent. The conclusion made was that bees between four and eight days old are usually nurses of older larvae. All the other stages of a bee's life, from birth to death, were studied in a similar way.

It should be pointed out here that the first bees of the season hatched out of eggs laid by the queen at the end of winter—in February—differ from bees hatched out of eggs laid in autumn with regard to the period at which they perform their duties. We may well suppose that there are differences within narrower limits and that, for instance, May bees differ in this respect from July bees, as do bees in different years, in different localities, and different stocks.

And this is quite normal, for the terms of performing the social duties are regulated by the needs of the community while the mobility of the limits of each group and the ability of bees to perform several (not all) kinds of work make the whole system adaptable to the changing conditions. Lately the ability of bees to perform different duties at one and the same stage of life has been an object of close study which has revealed that bees can perform even the functions of nurses of the older and younger larvae simultaneously. The general conclusions drawn from this study are highly interesting. While L. I. Perepelova studied normal behaviour of bees and discovered the laws governing the change of duties in connexion with a bee's age, M. Lindauer studied the exceptions, and discovered that there are individuals which linger at some or other stage of development, thus specializing in the performance of the duty appropriate to that stage. They perform the duty of, say, builder or guard for an exceptionally long time and easily go back to that stage. One can easily see that such bees are more diligent and adroit in performing their particular duty than others, so that research-workers studying this problem speak of the "inclinations" or even the "physiological predisposition" of individual bees. But this in no way contradicts the facts cited above.

When bees of some age group or other find no work to perform, they become motionless on the combs, thus economizing both their energies and food needed for movement. At the time of the main honey-flow, however, young bees below "field age" join the ranks of the foragers, thus entering an older age group before their time. It looks as if field-bees were recruited from younger age groups.

When, on the contrary, a hot wind and a scorching sun dry up the nectar in the flowers, the honey-flow is interrupted and the attractive smell of honey and wax starts spreading from the hives, both young bees of the preguard-duty ages and old foragers become active guards of the colony's wealth. A reinforced guard is posted at the entrance and on the landing-board. Each bee can see in all directions without turning her head and is ready to take wing and repulse an enemy attack.

FIELD DUTIES

Young Bee's Baptism of Air. Flight Intensity and the Landing-Board. An Automatic Checker of Goings and Comings. Bees and Electricity. When the Weather Prevents Bees from Flying. How Many Bees Die within the Hive? Flight Time and the Number of Forage Trips of a Bee.

While performing the cycle of duties within the hive, a bee at the same time prepares for field work. These preparations take her more or less time, depending on conditions, and seem to begin with short exercises on open cells which are soon followed by short flights quite unlike the flights of the foragers. On a warm and windless afternoon a great number of young bees may be observed taking their "play flight" as it is called.

A young bee hovers a little distance in front of the entrance at no great height and always facing it. After two or three minutes in the air she alights and goes back to the combs to resume her duties of cleaner or nurse. On specially fine days such flights are repeated, the bee again flying in the vicinity of the hive with her eyes fixed on it. The Tula experimenters noted as an extraordinary event that Bee No. 15 on her fifth day flew several yards away from the hive.

After her "baptism of air" a bee increases the range of flight and stays longer in the air. Though she still faces the hive and keeps her eyes on the entrance, the bee feels more at her ease in the air. Bees of eight or nine days of age fly around the hive in little and merry packs, looking at it from all sides, and return home some five to seven minutes after leaving it. This is their first long orientation flight. It is thought that in this way young bees study the locality and its salient points and familiarize themselves with the surroundings of the hive.

On the same or the following day, the bee is transferred into the last class of the "flying school": she may stay in the air as long as half an hour and fly without having to keep the hive in view. Like seasoned forager, she flies straight from the entrance along the air route—to the meadows, the woods, and the orchards. There, she alights on a chosen flower and spends some time resting; she does not yet attempt to collect nectar or pack her pollen-baskets—she merely flies from one flower to another and again rests before starting on her way home. Once back in the hive, the bee rests once more, then eats some honey before leaving home for the second time.

While many bees make several such idle flights before filling their crops with nectar and loading their pollen-baskets, some have been observed to master the art of foraging much quicker--after one or two trial flights.

There are records of individual bees, which, on first leaving the hive and hovering about it for some time with their heads towards it, did not return home but flew in ever expanding circles, thus

Many people believe that a bee entering on field duties brings into the hive loads of pollen, and not of nectar

making the play flight an orientation flight. Then they disappeared and some thirty or forty minutes later returned with their first loads of nectar or pollen.

This is how a bee, which a fortnight or three weeks ago gnawed her way out of the cell and first saw the light of day, becomes a seasoned forager. By this time the sugar content in her haemolymph has almost trebled, which is natural, since sugar is the fuel which a field-bee needs in greater quantities because her expenditure of energy is much greater than that of a home-bee.

Field-bees possess, among other qualities, the ability to feel the coming change of weather. Although much overrated in old bee-literature, this ability does exist and it is quite true that on the approach of a storm cloud the foragers return home. At such a time the bee-keeper can observe a curious sight: the air is alive with hundreds of thousands of bees, flying pell-mell at various heights all heading for home through the narrow spaces between the trees surrounding the apiary. It should be remarked here that the bees hastening home at the approach of a thunder-storm have never seen one, so their personal experience cannot have taught them to expect lightning, thunder and rain from the cloud. And yet, no sooner does the sky become overcast and the cloud hide the sun than the bees leave the flowers and rush home.

Before each trip afield a bee cleans her antennae by drawing them through her antenna-cleaners (left) and the proboscis, which she combs with the brushes on her legs (right)

For some time the flying throng speeds noiselessly overhead, then breaks up over the apiary and quickly dissolves. By the time the first heavy drops of the thunder-storm shower fall on the ground the bees have all disappeared. Deep within the entrances the guard-bees are hiding, now and again one of them appears, moves her feelers suspiciously, and instantly hides from the rain.

Surprising differences may sometimes be observed in the flying habits of field-bees.

In an experimental apiary there was a colony which was distinguished for exceptional order in the movement of incoming and outgoing bees. It often happens at the busy hours of the honey-flow that bees going out of the hive run into home-coming bees, and bees alighting with their loads collide with bees taking wing. Whatever other conditions may be, this certainly weakens the field activity of a colony.

The bees in that particular colony left the hive at the top of the entrance, ran along the front wall and took wing off it while the bottom of the entrance was left for the passage of the loaded foragers which moved into the hive from the landing-board. Thus, traffic in that hive proceeded in two rows as it were, the upper row consisting of outgoing bees and the lower, of incoming ones. This detail is rather important, if we take into consideration how congested the landing-board in front of the hive becomes at the height of the honey-flow.

When an attempt was made to study in detail the flying activities of bees it proved too much for individual observers or even groups of observers. When the honey-flow is at its height the number of bees going and coming in a strong colony is so great that they cannot be counted at sight.

To gain exact data as to flights in and out of the hive, a special apparatus had to be designed. After numerous experiments A. E. Lundie succeeded in constructing an apparatus automatically registering the passage of bees.

The entrance of a hive under experiment (the colony inhabiting it weighed 2.5 kilogrammes and numbered 25,000 bees) was blocked by thirty tiny automatic rocking levers, 15 millimetres long and 7 millimetres in diameter each. All bees whether leaving the hive or returning to it had to pass through thirty pipes or tunnels, half of which opened from the hive and let the bees out and the other half opened from the landing-board and let them in; these entrance- and exit-tunnels were placed side by side.

When a bee entered a tunnel, her weight caused it to drop; the little trap-door fell to and while the bee was passing no other bee could enter the tunnel. Dropping under the bee's weight, the tube touched a contact, closed the circuit and the electric current opened a door either into the hive or out, on the landing-board. The tunnel resumed its former position as soon as the bee left it to let in another bee.

The thirty tunnels were connected with thirty metres which automatically registered the number of passages every quarter of an hour.

The corridors in front of the exit-tunnels were equipped with a contrivance enabling the research workers to keep a separate account of dead bees carried out by the cleaners.

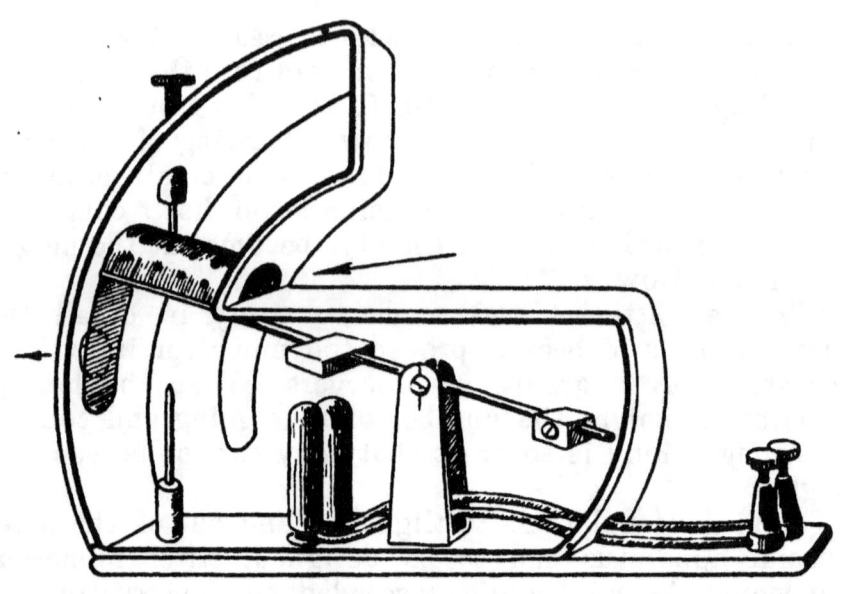

Thirty small automatic rocking levers put side by side were the most important part of the apparatus registering the passage of bees out of and into the hive. The one represented here is an exit tunnel: as the bee enters the tunnel, its weight causes the tube to drop and open the way out

Automatical apparatuses simultaneously registered the temperature of the atmosphere, the strength and direction of the wind, the atmospheric pressure, the intensity of sunlight, the quantity of precipitations, and the weight of the hive.

Everything was foreseen and provided for but, as usual, practical experience suggested additions and improvements.

Electrochemical corrosion of the copper plates in the apparatuses changed their weight; the tunnels constructed for the passage of one bee proved too small for bees with big loads of pollen in their pollen-baskets: foragers with overloaded baskets lost part of their burden and the bees coming after these lingered in the tunnel to pick the pollen up.

Peep-holes had to be made in the apparatus in order to establish causes of lingering, but bees tried to get into them through the glass, and spent still more time in the tunnels. The glass of the peep-holes had to be painted over. Then electric light by which the apparatus was inspected at night attracted the bees, a circumstance which caused additional confusion in the data.

There were many other failings in the work of the apparatus, although it had been constructed with much care and every detail in it had been well thought-out.

At first the bees felt the weak current in the contacts and retreated before the unseen foe, but finding no safer and more convenient way, they returned and overcame the invisible barrier. The more daring among the bees attempted to sting the rubber insulation of the wires, from which spread an intangible and yet quite perceptible force. But the rubber did not mind being stung and bees resorted to another means of defence; they started covering the insides of the tunnels with propolis. The experimenters had to improve the insulation, and at last everything was arranged satisfactorily so that the experiments could be started.

Operating from morning till night for 105 days the apparatuses registered 2,434,666 goings and 2,357,769 comings, nearly 5,000,000 flights in all. The total weight of the bees thus registered was half a ton.

In this experiment the research workers wanted to study the influence of the sources of nectar, of the time of day, and of the weather, on the intensity of field activities; they wanted to establish the duration of trips under different conditions, the number of trips a day, the mortality rate within the hive and outside, etc.

The apparatuses showed that bees have their flying and their non-flying weather, and established the velocity of the wind at which bees are grounded. Wind with a velocity of 5 metres a second considerably reduced the number of trips. The influence of temperature was less obvious: in early spring the bees made forage trips at 12 to 14°C., while by the end of May field work was started at 16 to 18°. Evening flights depended more on light than on the temperature outside.

When, however, the showings of the photometer and thermometer were compared with the blossoming time of various honey-plants and the secretion of nectar in flowers; it was found that field flights had been determined mainly by the field, pasture, food.

A trap set under the tunnels, into which all scavenger-bees fell when removing dead bees from the hive enabled the researchers to count them and to establish the fact that

Black, hairless and with frayed wings, an old worker-bee preserves the urge to pass her load of nectar to her sisters to her last hour

an overwhelming majority of bees died outside the hive; the number of bees dying at home was one or two to a hundred.

Of course the apparatuses could not reveal all the details of this phenomenon. They recorded the passage of both a vigorous field-bee on her foraging trip and of a dying bee which passed outside only to crawl as far from the hive as possible.

It has long been known that, like other animals, bees and the old queen, too, avoid dying in the nest. This instinct spares the colony the necessity of removing dead bees.

In the colony under experiment, bees died their natural deaths after making an average of thirty-two trips, which is an average of twenty-one hours in the air. In stronger colonies an individual bee is, naturally, several times more efficient: she makes more trips at shorter intervals.

But time has come for the bee to die. She is darkened with age and no hairs are left on her body. With a great effort the bee reaches her hive on her worn-out wings that have carried her for many miles and passes her load of nectar to a home-bee. Very old bees can collect only nectar, for no pollen can stick to their smooth chitinous bodies no longer covered with fuzz. Having left her burden in the hive, the bee drags herself outside. Stumbling, falling on her side and painfully rising on her legs, the bee crawls to the end of the landing-board, and, with the last breath of life, half falls and half flies to the ground to die there and thus render the last service to her colony.

SIMILARITY AND DISSIMILARITY

Why Are Workers Unlike the Queen and the Drone?
What V. I. Michurin Proved by Creating the Kandil-
Kitaika Apple. Nurse-Bees and Michurin's Mentor
Method. Two Figures Expressing the Bee's Life Story.
How Workers Resemble Queen and Drone?

Now that we have followed the life span of the worker we may well ask: How is it that the queen and the drone produce offspring capable of performing a wide variety of operations which neither of them can perform? We know that the queen spends practically all her life in the hive doing nothing but lay eggs, that the drone does nothing in the hive and merely awaits a meeting with a virgin queen; neither the queen nor the drone has wax glands; neither of them can build combs or has ever built a cell; neither has tools for gathering pollen or nectar; neither is capable of storing food or bringing into the hive a single drop of nectar, a single load of pollen; neither is capable of feeding brood. A well-known saying maintains that children resemble their parents, but then why are worker-bees, the offspring of the queen and the drone, so unlike their parents? From whom do the workers inherit their talents, since neither the father nor the mother possesses a single one of them? What conditions determine this dissimilarity between parents and children?

We remember the example of black-blood represented by three bisexual forms one of which is more like male, the second more female, and the third an intermediate between the two. After pollination, each of these plants produces offspring of all the three forms and thus part of the progeny is naturally unlike the parents. But in this case the three forms are fertile and, consequently, capable of propagating their kind by more or less complex means. With bees, however, the third form is sterile. Where does it come from? How is it reproduced in the progeny of the queen and the drone? Pondering upon these questions, Darwin justly said that the example of social insects and hive-bees in particular was a special difficulty in his theory and that it might "well be asked how is it possible to reconcile this case with the theory of natural selection?"

To elucidate this confusing phenomenon, Darwin cites the example of the annual stock, in which he sees a similarity with worker-bees.

The stock is a well-known decorative plant from the seeds of which grow thick tufts bearing double flowers of different colours—snow-white, cream, lilac, pink or spotted (mottled). The double flower is made up of petals alone and contains no stamens or pistil; it withers without producing seed. But whence come the seeds for propagating the variety? The seeds are produced by the few homely single plants little suited to adorn the flower-bed.

Through long and careful selection flower-lovers have bred the fertile varieties of the stock producing seedlings with beautiful but sterile double flowers and with some single and fertile plants. Darwin's conclusion is that such a variety could have been created only by selection having been applied to the family, and not to the individual.

If we draw an analogy between the flowers of double stock and the sterile worker-bees on the one hand, and between the pistil and stamens in the single stock plants and queen and drone, on the other, then Darwin's explanation will help us to visualize the possible way of the formation of the bee-species in which infertile progeny normally differs from the fertile parents.

"This difficulty, though appearing insuperable," Darwin writes, "is lessened, or, as I believe, disappears, when it is remembered that selection may be applied to the family, as well as to the individual, and may thus gain the desired end."

But this explanation was not exhaustive, since it shed light only on one aspect of the phenomenon and did not explain in what way habits peculiar to sterile female insects can influence the males and the fertile females which alone produce offspring. Idealist biologists instantly noticed this flaw in Darwin's great theory and were not slow in availing themselves of an opportunity of disputing the materialist teaching of Darwin.

Professor August Weismann, a German scientist, was the first to do this. The materialist theory of the development of living nature is based on the assumption that environment and conditions of existence play a decisive part in the acquisition of new characters and properties by an organism and considers the inheritance of these new characters and properties not only possible but necessary. But Weismann tried to prove that the properties and characters acquired by a plant during its life cannot be inherited and that this type of inheritance does not and cannot exist. To him, bees seemed an important, even a decisive, proof of his assertion.

Sterile forms cannot influence heredity, since they produce no offspring, Weismann exclaimed. We have "sexless" individuals in insect communities, which means that there are in nature animal forms incapable of reproduction but constantly reproduced by parents with whom they have no likeness. And yet, these animals, themselves incapable of transmitting anything to the offspring, have undergone modifications in the course of the earth's history.

"Does this not destroy the last stronghold of our opponents," Weismann gloated. "Is this not a proof that heredity is determined solely by a special hereditary substance which is never generated anew but constantly grows and multiplies, and for which the living body is merely a passive container, a nourishing medium, a case."

Weismann taught that the "bearer of heredity is contained in the chromosome material," and the chromosomes "represent a *separate world*, as it were, a world independent of the body of the organism and its conditions of life." He asserted that the cause of change in the heredity of organisms can lie only in the spontaneous change of the hereditary substance, that all such changes are accidental, indefinite, subject to no external influences, and that the environment, the external conditions act like a sieve in selecting out of the sum-total of all changes such as contribute most to the adaptability of the species.

The gist of this theory was, as we see, the invented hereditary substance, and the principle of Weismannism consisted in the unpredictability of changes in the hereditary substance.

Agnosticism denying the objective necessity of nature, the discouraging idea of unknowability—that is what lay at the bottom of Weismann's teaching.

Starting from his pseudo-scientific premises, Weismann discarded Darwin's idea that the infertile social insects had lost their fertility only as a result of other changes. According to Weismann, the prime cause of the appearance of infertile worker-bees could have been only accidental changes in the chromosomes, the hereditary substance.

Basing themselves on Weismann's principle of the "determinants," his followers diligently invented complex schemes of the action of the chromosome apparatus of heredity in bee life. In accordance with these schemes and putting forward the argument that females hatch from fertilized eggs laid by the queen and males from unfertilized eggs, some of the Weismannists went so far as to state that in the bee-colony the queen is the repository of only male determinants, owing to which her unfertilized eggs produce drones, and that drones, on the contrary, carry female determinants, which is the reason why female queens and workers develop from fertilized eggs.

Thus the false teaching that the embryo plasm is independent of the body logically led Weismannist-Morganists to the absurd conclusion that the father and the mother are not the parents of their children; that for their children the parents are brothers or sisters; that a female is merely a female covering of the male substance while a male is a deceptive exterior hiding the female substance.

Certain biologists studying the problem of the appearance of social insects, such as wasps, ants and honey-bees, tried to prove that the colony consists only of males and females, the majority of the latter—the so-called workers—being underdeveloped owing to undernourishment.

This seems convincing enough at the first blush, but can anyone who stops to think agree that the worker is merely an "underdeveloped" queen unable to realize her hereditary properties and characters owing to undernourishment? This premise, strictly speaking, is not far removed from Weismann's views.

Here are the brains of worker (centre), queen (left) and drone (right), equally enlarged

The development of a worker takes four to five days longer than that of a queen and her nervous system is better developed than that of the queen. How, then, can the worker be considered an *underdeveloped* insect?

Late in the nineteenth century, when biologists were arguing as to what conditions produce sexless workers in the bee-colony, I. V. Michurin who lived in a Russian town then called Kozlov, and whom the world did not know yet began his study of the influence of grafting on plant varieties.

Michurin studied the changes in heredity produced outside the sexual process, through changed conditions of life and changed food in plants, but it is clear now that the laws he discovered in the vegetable kingdom furnish the key to the secret of heredity in bee life as well.

Plant-lovers are well familiar with Cytisus Adami, or Adam's broom, a bizarre tree which is mentioned in many text-books. There are many descriptions of its dull red, bright yellow, and purple flowers growing on one tree with differently set branches covered with sharply differing leaves. This plant often has flowers of two different kinds in one cluster, and sometimes one flower is divided into two, one half being yellow and the other purple. The yellow half of the dorsal standard is bigger and the purple smaller. There are even flowers with a bright yellow corolla and half of the sepals purple.

This tree, a living and growing mixture of the common yellow and purple brooms, was obtained by Adam not through crossing but by grafting on to ordinary broom a bud from the beautiful bush "golden shower." After a year's rest the grafted bud broke into many shoots, out of which the most vigorous was chosen and propagated. From this shoot sprang the varicoloured Adam's broom in which the tissues of two different varieties have blended without crossing and the resulting plant produces offspring with leaves and flowers of a pronouncedly hybrid origin, intermediate between the stock and the scion. Here heredity was changed obviously and solely through a change in food.

Darwin considered this "a most important fact, which will sooner or later change the views held by physiologists with respect to sexual reproduction," and his foresight was corroborated seventy years later, in 1938, when in an introduction to the complete works of I. V. Michurin, Academician Lysenko for the first time clearly formulated these new views with respect to sexual reproduction by stating that: "Assimilation also takes place when two sex cells fuse."

This exceedingly clear statement shows that hybrids obtained through successful grafting, as in the case of Adam's broom, do not differ in principle from hybrids obtained through usual crossing.

But the broom was an unexpected and incomprehensible freak in the practical experience of a gardener, the work of blind hazard, and it was not Adam who enriched biology with a very important discovery.

It fell to I. V. Michurin's lot to demonstrate by breeding late in the 19th century a new variety of apple-tree—Kandil-Kitaika—that grafting and fusion can be made the means of not an unexpected and chance, but of a conscious and purposeful influence on a plant. In Michurin's selectionist work there appeared plants (he called them mentors) which fed with their sap hybrid scions; the latter being hybrids were more pliable, more apt to be influenced by the nature of their mentors through which some of their properties were strengthened, others weakened, and still others altered.

Bee-pollination increases the crops of fruits and berries several times over

The Weismannist-Morganists, votaries of idealist pseudo-science, refused to accept the idea that the heredity of plants can in any way be changed by this method, but with the help of the mentor method I. V. Michurin produced scores of new varieties and proved that feeding fosters natural characteristics. He continued to use his method and created a large number of new fruit varieties which made his name famous. Michurin's pupils have further elaborated the mentor theory and perfected the technique of applying this potent method for changing the nature of plants.

In his closing speech at the session of the U.S.S.R. Lenin Academy of Agricultural Sciences held in August 1948, Academician Lysenko summed up the long argument of the Michurinists against the Weismannists: he demonstrated a tomato plant with red and yellow fruits growing together in one cluster.

The branches with differently-coloured fruits were cut from plants grown from seeds. But the seeds had been obtained *not* through cross-pollination of the flowers, but from a hybrid obtained through the blending of a red-fruit and a yellow-fruit tomato plants; they were vegetative hybrids; the seeds were from an offspring of a variety whose nature was altered by the mentor method. This was a living demonstration of Lysenko's words that "heredity is determined by the specific type of metabolism. You need but change the type of metabolism in a living body to bring about a change in heredity."

Now let us recall the process of rearing the worker and the queen, let us recall laying workers and queens with worker characteristics, and the feeding of larvae, described in foregoing chapters and showing what fundamental changes in the nature of the bee, in her anatomy and instincts are brought about by a change of diet, and ask ourselves: are not the laws obtaining here similar to the influence of the mentor in plants nurtured on changed food?

In this case bee biology furnishes a tangible example of nature directing the development of organisms through feeding, through directed metabolism] in developing embryos—the larvae.

This is not an abstruse hypothesis. The most experienced of bee-keepers, who know their profession down to the finest details, have long been utilizing the influence on the nature of queens of the nurse-bees that rear them and—through them—improving the colonies.

Special experiments have proved the influence exercised by nurse-bees on their feedlings.

At the Tula Experimental Station, A. S. Mikhailov removed combs containing eggs of a strain of bees with long proboscides and placed them in a colony of bees with short proboscides, and out of the eggs laid by the queen with a long proboscis hatched bees with shorter proboscides. He reversed the experiment and placed combs with the eggs of a short-proboscided queen into the hive of bees with long proboscides. The bees hatched from these eggs had longer proboscides.

And yet the results of A. S. Mikhailov's experiments were slow in gaining recognition. In this they shared the fate of the assertions of many practical bee-keepers that nurse-bees are capable of considerably changing the heredity of the brood they feed. So it became necessary to clarify this problem, the significance of which is great both for the theory and practice of bee-keeping.

* * *

In the summer of 1949 at the apiary in Gorki Leninskiye, near Moscow, and at the apiary of the Central Experimental Station in Barybino, also near Moscow, were conducted two parallel series of experiments studying the role of nurse-bees. The aim of the experiments was to reveal whether royal jelly secreted by nurse-bees could influence heredity.

It was decided to plan the experiments so as to bring out the influence of nurse-bees not on anatomic characteristics which are more ancient and more stable, but on behaviour, which is more susceptible to influences. At the same time it was clear that to come to correct conclusions, such characteristics of behaviour as, for instance, "fussiness" would not be sufficient. Although all bee-keepers know that bees may differ in respect to this quality, the danger of incorrectly assessing expressions of it was too great, since it would be difficult to establish an objective criterion.

To find a clear answer to the question, it was decided to study the influence of nurse-bees on an obvious and indisputable varietal characteristic, such as the capping of honey-filled cells.

As has been noted above, black forest bees from the north seal each cell containing honey with a white convex cap slightly *above* the level of honey in the cell leaving a little layer of air. Grey mountain bees from the south, on the contrary, seal their cells with a flat corrugated cap placed *directly on* the honey so that the cap seems dark and wet.

The characteristics of the varietal behaviour of bees seem to be materialized and concretized in the type of capping. It would have been hard to find a better object for the experimenters in the solution of the task before them.

Combs sealed by southern bees were put into a hive with bees that had been reared by northern nurses from eggs of a grey mountain queen. The combs were filled with honey from top to bottom and sealed with "wet" corrugated cappings, some of which were scratched away by the experimenters spelling the word "Food." Honey, like liquid gold, was flowing from the opened cells. The experimenters awaited with interest what the bees would do with the cells. For do something they must—instinct makes bees repair damaged combs and seal the cells anew. How would the daughters of the "dark-sealing" race reared by "white-sealing" nurses seal the cells?

The combs were in a glass observation hive and with every day the word "food" stood out more and more clearly in slightly raised white characters against the background of darker "wet" surface of the southerners' seals.

The experiment was repeated several times, and by the end of the summer a collection of combs accumulated on the laboratory desk with clearly-perceptible words "Food", "Variety," etc., wrought out in white cappings.

When K. A. Timiryazev experimented to prove that the formation of chlorophyll in a leaf is due to light, he covered a box of young cress seedlings with cardboard in which letters were cut out, thereby making sun-rays "write." And in the bright green of the lighted plants against a yellow background of colourless shoots the sun wrote the word "Light." Timiryazev called this experiment "a photograph of life."

In the experiments we are describing the combs confirmed by cell cappings that the formation of a variety is connected with the action of food.

Just as the Michurinists changed through grafting the nature of the yellow-fruit tomato variety into a red-fruit one, the bee-experimenters, through a change in food, changed a dark-sealing strain of bees into white-sealing, thereby convincingly proving that a natural "mentor" plays a very important part in shaping the heredity of a bee-colony.

The rearing of the brood, highly perfected through selection, has become a biological property of the bee. It provides a further demonstration of the correctness of the Michurinists' conclusions regarding the shaping of heredity, and helps us to understand how it is possible for a sterile worker-bee feeding the larvae with the secretion of her brood-food glands to reproduce herself in succeeding bee generations, at the same time transmitting to them the changes in anatomy or in instincts acquired by the colony.

Now let us reconstruct the life-cycle of the worker-bee. After emerging from the cell as a mature insect she lives about six weeks. Here we mean only summer bees.

People usually think of the bee as a creature that spends its life in flight, moving over flowers, bathing in the flood of sun-rays, sucking sweet nectar and, covered with golden pollen, imbibing the heady scents of spring-tide.

This idea is naïve and very far from the truth. Out of her six-week life a worker leaves the hive for a few dozen hours, almost 900 summer hours out of a thousand being spent in the hive. Autumn bees, which live as long as 5,000 hours, spend in the hive something like 4,900 hours, and for several long months form part of the ever-moving cluster in which the colony finds protection from winter frosts.

The twilight and warmth of the hive—these are the worker-bee's element, these make up the environment of her life.

The few hours spent in flight are only fleeting episodes in her life, they are only sun-lighted intervals in the constant darkness of the hive. In the behaviour and ways of the worker (so unlike either of her parents) we see a queer blending of the queen's devotion to her home and the roaming spirit of the drone.

Of course, calling this a likeness is a crude way of speaking, here we see but an embryo of a likeness. But then the queen and the drone reproduce only the embryo of the worker-bee, the rearing and upbringing of this embryo is the occupation of the workers, who, through food, can cause the worker larva to depart from both the parents and, through royal jelly and brood food of the nurse-bees, can implant in it their own characteristics and instincts.

The direct and indirect influences of the environment assimilated by the workers are transmitted with food through one channel to the laying queen in whose body this food is transformed into an egg, and through another to the larva hatched from that egg and producing a worker, a drone, or a queen, according to the conditions of the development of each egg.

That is how every individual, and in the final analysis, the whole colony is reared and transformed. In thousands of interwoven individual life-cycles the colony develops spiral-like, each turn of the spiral reproducing the road travelled by the progenitors. At the same time the colony itself goes its way, which, of necessity, will be followed by its successors.

THE WAY TO NECTAR

THE COMMUNITY'S BREAD-WINNER

Bees on Shepherd's Purse and on Hazel Catkins. How
a Kilogramme of Honey Is Gathered. The Direction of
Flights and Fixations. Time Factors in Working Flowers. Mixture of Pollen in Loads Brought to the Hive.

In order duly to appreciate the thoroughness, persistence and pertinacity with which the winged pollinators
work, one should observe a bee on shepherd's purse. This
cruciferous plant is not included in any list of honey-
bearing plants, for, although there is plenty of nectar in
its fragrant flowers, the flowers themselves are extremely
small, homely, and sometimes all but imperceptible.
Bees on shepherd's purse is a sure sign of a poor
honey-flow.

To collect nectar or pollen, a bee must be comfortably
seated on the flower. In the flowers of the plants she is to
pollinate the bee finds not only comfortable landings but
a convenient arrangement of nectaries, stamens and stigmas, and even coloured markings in the shape of lines or
dots—the so-called honey signs—showing the way to the
nectaries. But the flower of shepherd's purse affords the
bee no footing whatever: the bee is here a chance visitor,
and no conveniences are provided for her. The small size
of the corolla set on a long thin pedicle and the arrangement of florets in the cluster make it difficult for the bee
to alight. And to crown all this, the bee is often too heavy
for the pliable stem of the plant, so that when she clings
with all her six legs at the pedicle on top of the stem and
inserts her proboscis into the calyx, the stem preserves its

balance for but a short moment. This, however, is sufficient for the bee to catch a firm hold of the flower. Under
the weight of the bee the stem then bends low and measures its length on the ground. But the bee, falling together
with the flower and lying on her side or upside down, with
her back or even head to the ground, continues lapping
with her spoon the nectar hidden in the flower which is
several times smaller than herself. The bee leaves the flower
only after sucking it dry of nectar; she then flies some
distance from the flower and awaits in the air until the
stem, freed from its burden, straightens up. Then she alights
on the next flower on the same stem and, clinging firmly
to it, starts taking the nectar, the stem again bowing to
the ground.

Watching the persistence with which the bee works
shepherd's purse, we must remember that from one flower
she can take mere hundredths of a milligramme of nectar,
a drop as big as a pin-point. The bee has to repeat her
acrobatic feats on shepherd's purse dozens of times before
her honey-crop is partly filled with nectar.

But this is an extreme case. On more "convenient"
flowers nectar is collected quickly and methodically.
Here the general and particular anatomic correlations in
the structure of both the flower and the bee are revealed
with especial clarity. With precise movements of each
part of her body the bee inserts her proboscis into the nectaries to see if they contain nectar. These movements
are performed from different positions for different flowers,
the most convenient position generally being adopted. If
the nectary is empty the bee flies to another flower, if
it contains nectar she sucks it up. Having worked one
flower, the bee goes to the one next to it.

It is also interesting to watch the behaviour of pollen-
gatherers.

Dandelion pollen is humid and sticky, and sometimes
bees roll over the flower, brushing their sides against it,
to cover their hairy bodies with pollen grains which later
on they remove with the combs and brushes on their legs
and then put into the pollen-baskets.

On reaching the landing-board of her hive with a load
of the humid and heavy pollen, the forager rests for a con-

siderable time, now and then making the fanning movements. This she does to dry the pollen.

The behaviour of bees on alder or hazel catkins is entirely different. Both alder and hazel produce dry, powdery pollen. A very slight shock will cause the flowers to eject all the ripe pollen grains. And the bee flies to the catkin from below, cautiously alights on one side and slowly and carefully proceeds upwards. If a pollen grain falls, it will be arrested by the hairs on the bee's body.

Now and then the bee hangs from the catkin by one foreleg, removing the pollen from head and abdomen with the other and passing it to the middle pair of legs.

A forager buries herself in the corolla of a dandelion, where she finds both pollen and nectar

A bee clutching with all her legs at a shepherd's purse flower on a drooping stem offers an illustration of the foragers' pertinacity, while a bee catching with one claw at a hazel catkin shows how proficient they are at gathering food.

The movements of a bee working a flower are so quick that it is impossible to analyse them with the naked eye; only velocity filming has enabled us to see in detail how a bee collects her load of pollen.

Taking-off, landing, burrowing among the anthers and the quick movements of the bee's legs all seem merged into one. But if we observe the sequence of movements, which can be done more easily in an open flower like the poppy, wild rose or apple-blossom, we can see the bee, upon alighting in the centre of the corolla, vigorously scratch pollen grains off the anthers with her mandibles and moisten them with nectar; soon she is covered with pollen grains sticking to the hairs on the head and the thorax. The bee repeatedly rubs her body with her fore and middle legs, passes the antennae through the antenna-cleaners, cleans the proboscis, eyes, and thorax, and rubs one leg against the other, all the time bustling among the anthers. Pollen collects on the brushes of the middle legs and is now and then combed off by the spines on the hind legs, which also brush the pollen from the body.

Flying in the air and soaring, the bee continues to brush pollen off her body and gradually transfers the pellets of sticky pollen to the pollen-baskets on the hind legs. Under natural conditions all this happens in a much shorter time than it takes us to describe it.

The chain of movements resulting in filling the pollen-baskets is repeated several times and while the hind legs are completing one series of operations the fore legs have started the following one. This explains why all the movements seem to occur simultaneously.

In working blossoms producing dry, powdery pollen, a forager ascends the catkin from below

In addition, during her flight from flower to flower the bee continues to press the two loads in her pollen-baskets. The loads in the baskets on the right and left legs are always of equal size and this is quite natural, for it would be difficult for an unevenly-loaded bee to reach her hive.

Bees on flowers deserve special attention.

All living creatures display persistence in obtaining food for themselves and for their offspring. The roots of plants find their way to moisture sometimes through layers of stone. A mountain goat will scale inaccessible cliffs and jump over precipices to reach a green bush. A sea-gull flies dozens of miles off to sea to catch a small fish for its young.

But when a field-bee leaves her hive she has already had her meal of honey, she is not hungry. She does not feed directly on nectar or pollen and she no longer feeds the brood herself.

Sic vos non vobis mellificatis, apes, Virgil said in his *Georgics,* although in his time it was not known that a forager may be dead by the time the nectar she collected is transformed into honey; it was not known then that a bee collects honey for the colony in which she will not live long, for the larvae which she will not feed.

A bee collects food for the community as a whole. No matter how much honey the hive contains, the bee will go on and on bringing more in as long as there is nectar in the flowers and empty cells to store it in.

A forager never tires of sucking and lapping food with her proboscis wherever she can find it,

Dorothy Hodges, a British bee-keeper and artist, has made a series of drawings in which is artistically depicted the process of gathering pollen into the pollen-basket. She has explained the function of the central hair of the pollen-basket, forming the axis of the pollen load as it were. Her drawings clearly show how the hairs on the tibia hold the load from outside

but she does not satisfy her hunger in the process, she does not eat what she finds.

Similarly, bees avidly sucking water do not drink, they do not quench their thirst in this way: as has been stated elsewhere, both the nectar and water a bee gathers go into the honey-crop with chitin-laid walls. Like the load of pollen in her pollen-baskets, liquid food, too, is brought into the hive and put into the combs as stored food for the whole of the colony. A forager's honey-crop is not her stomach, not an organ for assimilating food which an individual consumes, it is a reservoir for temporarily storing communal food and at the same time a retort for its primary processing.

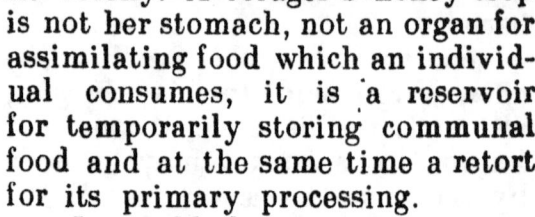

It would be wrong, then, to think that the bee's proboscis is her mouth. And true enough, the mouth proper with which the bee eats is a small folding muscle—the stomach-mouth — connecting the honey-crop with the alimentary canal.

This muscle is a cleverly constructed organ. It sucks up pollen grains that get into the honey-crop together with nectar and conveys them into the stomach proper. The valve can open to honey the way from the honey-crop into the alimentary canal, when necessary. The quantity of food it lets

In her vivid and exact drawings, D. Hodges shows in a dynamic form the movements of the legs with which a flying forager continues to distribute and pat down the pollen masses. From the drawings the details in the position of the hairs and spines on the bee's legs become quite clear

pass is just enough to enable the bee to perform her work. If the bee works a lot, cleaning the hive, feeding the brood, building combs, carrying water or nectar, then the muscle valve lets more food through. If, on the contrary, the bee is inactive in the hive, the valve relaxes and less food is consumed. In this way the anatomic structure of the bee is perfectly adapted for the satisfaction of the needs and requirements of both the individual and the colony.

If the colony has little food, all the bees in it get little, if it has enough, all the bees get enough. But if the colony has an excess of food *no bee is capable* of eating too much: the surplus is stored against a rainy day.

A forager emerges from a cell built by the preceding generations; she is nurtured

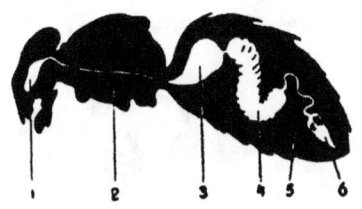

The alimentary canal of the bee along which food passes in the insect's body. 1. pharynx; 2. oesophagus; 3. honey-crop; 4. proventriculus; 5. small intestine; 6. rectum. The junction of the honey-stomach and the proventriculus is shown in detail in the figures on pages 184 and 352

with food brought into the hive by her elder sisters and she brings food to the hive not so much for herself as for her younger sisters, for the future generations.

For each bee the colony is nest, warmth, food, protection from enemies and the possibility to take part in the propagation of the species. All this means life and this is what each bee in her turn gives the colony.

We have mentioned above that before leaving the hive a field-bee takes some honey to be able to return if she finds no nectar in the flowers. Then we should bear in mind that a flying bee needs fifty times the amount of oxygen needed by a resting bee. The body temperature of a bee in flight is ten degrees higher than that of a motionless bee and a flying bee may be described as a warm-blooded animal. A certain quantity of food is necessary to generate the energy expended in flight. A study of carbohydrate metabolism in bees has shown that before leaving the hive a bee takes about 2 milligrammes of honey and spends about half a milligramme a kilometre. Thus this honey must last her about four or five kilometres. And that is exactly the distance bees usually cover.

A fielder brings home about 50 milligrammes of nectar which, upon condensing, yield twenty to thirty milligrammes of honey. From this microscopic quantity we must deduct the two milligrammes consumed by the bee, and the total the colony receives as a result of one flight is something like 20 milligrammes of honey.

The upper figure shows a lengthwise section of honey-stomach, proventriculus, small intestine and rectum. The x-shaped opening between four thick lips (stomach-mouth) at the summit of proventriculus projecting into the honey-stomach, is given below, still more enlarged

So we see that many thousands of "bee-flights" are needed to store in the combs a kilogramme of honey.

A kilogramme of the sweet food is the nectar from more than 100,000 dandelion corollas (each corolla consists of dozens of florets) or of from 1,500,000 to 2,000,000 acacia flowers, or from 4,000,000 to 5,000,000 sainfoin flowers, or from 6,000,000 to 7,000,000 flowers of red clover.

If we add up the distances covered by the foragers of a strong colony during a heavy honey-flow we get the equivalent of a flight to the moon. All the bees in a fair-sized apiary make during a summer the equivalent of several flights to the sun and back. This is no wonder, either, for it is known that the bees of one colony visit during the season nearly 250 million flowers.

In order really to understand the bee, we must always think of the many-mouthed colony of the tiny winged creatures. All this light and dynamic system seems to be tied in a knot of counteracting gravitations, the thousands of individuals that make it up occupying a certain volume of space in the air. The individual members of this system are in constant motion and, imbibing food drop by drop sometimes at a distance of several kilometres from the hive, bring in stores from everywhere.

Leaving their stationary hive and flying in all directions and at various heights, the foragers make their way into the remotest corners where they can find plants producing nectar and pollen. Then they return to their hive, leave their loads in the combs and once more scatter in all directions.

During a heavy honey-flow, the busiest time for bees, endless currents of home-coming bees meet the foragers hastening to the field. Thin dotted streams of honey flow from morning till dusk towards the narrow bee-entrance, beyond which the unloading and storing of honey is going on.

A loaded bee flies more slowly than an empty one

The bee's flight life is short and each moment of flight is expensive for the colony. That is why the bee has acquired through natural selection the instinct of economizing time and energy in flight to the maximum. This is easily seen even in cases when bees deviate from the proverbial "bee line."

In the memoirs of Kh. N. Abrikosov, an old Soviet apiarist, author of *Bee-Keeping in America*, there is an interesting story from the times when Abrikosov headed the big apiary of the Lesniye Polyany State Farm.

"I often observed," says Abrikosov, "that on calm days the bees flew over the tall pines to a buckwheat field situated in the middle of a forest. Through field-glasses you could see them there like sparkling gold specks. But as soon as the weather changed and a wind rose the bees took a roundabout way—along the forest road and a cutting. This observation was verified several times and showed that early in the morning and on windy days the bees did not try to fly over the forest but invariably took the road and the cutting. It looked as if the early scouts had brought in the news that the day was windy and that the roundabout way should be taken, which the entire forager force followed."

Bee-literature abounds in similar observations.

During the honey-flow, neither wind nor high water will stop bees flying afield. In his poem *Bees* the poet Nekrasov described an apiary on a knoll surrounded by water on all sides. The bees continued flying over the flooded area to the wood and to the distant meadows. It was easy for empty bees to cross the watery expanse, but on the homeward flight, loaded bees fell into the water so that it became alive with their struggling bodies. On the advice of a passing peasant, big branches were stuck into the bottom along the bees' route; home-coming bees landed on them to rest and safely continued their way home.

At this busy time no bee capable of going afield will remain on the combs, neither will she waste time while working the flowers. One more important feature of bee behaviour—a preference for certain flowers—becomes especially noticeable at the time of a copious honey-flow.

It was noted long ago that, although generally bees visit hundreds of various plant species, they collect nectar from the flowers of only one particular species during one flight, in which they differ from many other insects.

Acknowledging that "bees are good botanists" and unerringly distinguish species when visiting flowers, Darwin offered the following explanation of this phenomenon: "No one will suppose that insects act in this manner for the good of the plant. The cause probably lies in insects being thus enabled to work quicker; they have just learnt how to stand in the best position on the flower, and how far and in what direction to insert their proboscides."

And indeed the time study of bees' work on the flowers of various species has shown that a bee spends much more time collecting nectar on a flower she is visiting for the first time and whose structure she does not know, than during her following visits to similar flowers, after she has familiarized herself with them and learnt to work quickly all the nectaries. On the flowers of one variety bees noticeably economize time also in collecting pollen and filling their pollen-baskets with it.

This fixation instinct makes the bee the surest and most reliable pollinator for large-scale agriculture where extensive areas are sown to uniform crops of one particular variety with thousands and even millions of simultaneously opening blossoms of similar plants.

True, a microscope study of numerous samples of pollen from pollen-baskets of bees has shown that they contain, as has been stated above, an admixture of foreign pollen—material evidence that the bees have visited other varieties. More precise observations and analyses have demonstrated that the poorer the honey-flow the greater the number of different varieties and even species visited by bees. But we know now that this does not detract from the bee's value as a pollinating agent. The Michurin law of the importance of pollen mixture explains to us why it is not only harmless but even advantageous for the successful pollination of plants that bees' fixation habit exists side by side with certain deviations from it.

BEE "DANCES"

Do Bees Know Their Way to the Flowers? Scouts in Flight. Two Types of Dance on the Combs. The "Figure Eight" Dance and Its "Steps." The Solar Angle of the Air Triangle or Why a Bee Can Find Her Way Alone.

No sooner has a rich source of nectar appeared even in a far-off meadow than thousands of bees but yesterday inactive on the combs fly in an endless stream to exactly this spot.

How does news from the plant world reach the hive? Who brings word about the state of the nectaries in flowers to the bee-colony?

To find a reply to this question, a simple but convincing experiment was made a hundred years ago. Close by two beehives in a stone wall there was a niche closed by railings with creepers clinging to them. A saucer with slightly moistened sugar was put on a stool in the niche. A bee from hive No. 1. inhabited by yellow bees was brought on to the saucer. She sucked sugar for some time, then made a few circles over the saucer as if taking bearings, and making her way from the niche through the creepers returned to her hive. A quarter of an hour later about thirty yellow bees were flying about the niche as if looking for an entrance. One bee after another got through several layers of the living curtain and settled on the saucer. During the following days while the sugar was left in the niche it was visited by the yellow bees but not by a single bee from the neighbouring hive of black bees.

It was quite clear that the news of the sweet lure in the niche covered by creepers became quickly known to the colony of yellow bees and remained a secret for the black ones.

But how does a bee that has discovered a rich store of food tell others about her find? And how do the numerous new bees find their way to the spot discovered by their sister? It cannot be supposed that hive-bees learn of the blossoming of nectariferous plants within the range of their flight just by hazard.

It is impossible that blind hazard should be the determining factor in any sphere of life, in any of its manifestations. The world we live in and of which we form part is a world of matter developing according to laws; it is, as J. V. Stalin said, "a connected and integral whole, in which things, phenomena are organically connected with, dependent on, and determined by, each other."

Science, the enemy of hazard, had to reveal the interdependence and connections between flowers and bees.

The foragers fly with such certainty to the source of nectar as if they already knew the road.

In a chapter of *Anna Karenina* describing Levin's early morning sporting trip Leo Tolstoy writes: "... the minutest sound could be heard in the calm stillness of the morning. A bee buzzed past his ear. He looked up and saw a second and a third. They came out of a yard and were flying over the hemp field straight in the direction of the swamp."

A keen observer of nature that Tolstoy was, he once more noted this peculiarity of bee flight in describing the bee-garden Levin went to for fresh honey.

"In front of the openings of the hives, it made his eyes giddy to watch the bees and drones whirling round and round about the same spot, while among them the worker-bees flew in and out with spoils or in search of them, always in the same direction into the wood to the flowering lime-trees and back to the hives."

And indeed, bees seem to fly all in one and the same direction, one after another.

Moreover, each forage area is visited by exactly as many bees as can profitably work it.

In an experiment carried out in a locality where no honey plants grew, ten niterbush flowers were placed in jars of water at some distance from a hive. Five bees came to the flowers and were marked. After a while the same bees were seen working the flowers. The next day these five bees were again observed, four of them gathering nectar and the fifth pollen. Other bees were seen flying past the flowers but for some reason they did not alight on them.

Then the number of flowers in the jars was doubled and the number of bees working them increased to eleven, two of which collected pollen. The eleven bees continued to work the niterbush and no more came. And again other bees passed by the flowers and made no attempt to work them.

All this looked as if an experienced someone were directing the bees to this or that area in accordance with the volume of work there.

To study the problem in more detail, further interesting experiments were staged. Numerous observations invariably showed that bees are far more diligent and persistent in visiting large clumps of plants than solitary plants of the same species growing wide apart. This fact reveals one more simple and effective means through which nature encourages the cohabitation of masses of homogeneous plants and leaves unserved plants separated from the mass of their kind.

Still the question remains: How do bees locate forage areas and how do they regulate the number of nectar-and pollen-collectors?

Long ago bee-keepers suspected the existence of scout-bees in the bee-colony. Perhaps it was these scouts Alexander Pushkin spoke of in one of his poems:

> *As soon as the early flowerets*
> *Appeared where the snow had vanished,*
> *From its wonderful waxen kingdom,*
> *From its cell with the scent of honey*
> *The first little bee came flying;*
> *It flew to those early flowerets*
> *To gather sweet Spring's glad tidings. . . .*

But it is not only in spring that the scouts fly abroad. Observations show that a number of field-bees in a colony are regularly engaged in checking the state of flowers and establishing whether they contain nectar.

What are these bees? An answer to this question has been furnished by some of the experiments conducted at the apiary in Gorki Leninskiye.

It was noticed here that during the manipulations in the evening or at night the reaction of certain bees to the light of the lantern at the glass wall was extremely lively. While the rest of the population of the hive continued bustling over the combs paying no attention to what was going on around, certain bees—very few in number—hastened to the light and if it was moved, followed it like a magnet.

These light-sensitive bees were made to follow the lantern into the glass passage in front of the entrance and after being marked allowed to return into the hive. When observations were resumed at the bee-entrance the next morning the marked bees were among the first to leave the hive.

So it was as good as proved that scout-bees are particularly attracted by light. We have had occasion to mention that the percentage of sugar in the haemolymph of field-bees is high, and it is quite probable that this physiological peculiarity determines the field-stage in the development of each bee, just as the greater development of the glands at certain stages of the bees' life determines the activities of nurse-bees or architects.

So far so good. Suppose the scouts that leave the hive first are the first to discover a new source of nectar. But the scouts alone cannot feed the whole colony.

Now, let us put a feeder with mint syrup some twenty-five metres northward from the hive and wait for the first bee to visit it. Let us mark her with a white dot and see what happens next. After the first bee has reached her hive, the number of bees coming for the syrup will immediately increase. While they are busy we shall mark all of them and when we have marked, say, fifty bees, let us put more feeders at the same distance from the hive at the three other points of the compass. The syrup in these feeders is of equal concentration but devoid of any smell whatever.

After this nothing will change: the bees will continue visiting the first feeder with the smell of mint.

Then we repeat the experiment and pour into the three new feeders the same syrup as in the first and observe certain changes in the behaviour of the bees. As before, the northern feeder will be visited both by marked bees and by new-comers, but in addition, bees will come to the three new feeders, and mainly unmarked ones. Each feeder will be visited by approximately as many bees as the first.

The conclusions from the two experiments are clear enough: first, the smell of the food was in some way made known to the bees recruited by the scouts, and secondly, the bee that came to the hive from the mint-smelling feeder conveyed to the recruits that they should look for food smelling of mint but did not indicate the direction where it was. In the two experiments all the feeders were placed at equal short distances from the hive. Perhaps this is of some importance and perhaps nothing of the kind will happen if the feeders are placed at different distances and farther from the hive?

A feeder of carnation-smelling syrup was put at 750 metres from a hive. About a score of bees that came there first were marked. Soon after they reached the hive, new, unmarked, foragers came to the feeder. All the unmarked bees were put into a cage, so that only the marked bees could come to the spot, collect the syrup and return home. This was done to prevent too many bees visiting the feeder, which would have been confusing for the observers.

Later on the feeder was taken away and about ten pieces of cloth smelling of carnation essence were scattered at different distances but in the same *direction*. Observers were posted at each lure to count bees visiting it. Within a period of an hour and a half—the time during which the observations were in progress—only four bees were registered on the lure at 75 metres from the hive; not a single bee came to the one at 200 metres; five came to the lure at 400 metres; seventeen to the one at 700 metres and as many as 300 were seen on the lure at 800 metres. The lure at 1,000 metres was visited by twelve bees, and those still farther removed from the hive by very few bees. In a word, the greater number of foragers came to the lures at the distances least differing from the original one. Since of all these bees only the twenty marked foragers had visited the feeder previously, there was no doubt that the recruits had in some way been informed of the *distance*.

The use of the six-frame observatory hive has enabled the observers to make major discoveries in the flying activities of bees. A fairly strong colony can live in such a hive and the behaviour of the bees in it is closest to the natural. A long glass lobby connects the bee-entrance with the outlet, owing to which the great numbers of bees coming from the field do not confuse the observer

But in what way? Karl von Frisch, an Austrian professor, spent twenty years of his life finding a reply to this question.

The glass walls of observation hives and the marking of bees long ago enabled students to learn what the hive's emissaries do on returning from a successful forage flight.

Returning with a heavy load a bee is visibly excited; she runs through the hive-entrance, makes her way to the upper part of the combs and stops in a crowd of bees. She regurgitates droplets of nectar from her honey-crop and home-bees suck it from her mouth and one after another carry the nectar to the cells for storing. Then the forager begins spinning about on the comb describing little circles, clock-wise or anti-clock-wise. These characteristic movements were called "dances" and described accurately enough in 1823, but only a century later—in 1923—their meaning became known.

The bee continues her "dance" for several seconds, sometimes about a minute, and some of the field-bees that had been doing nothing before she came join in the dance. They hastily follow the dancer, and seem to stroke her with their outstretched antennae, repeating her every movement (this is worth noting!).

The dancer then runs to another place on the combs and performs the quick steps of her dance before another group of bees, after which she makes another trip to the source of nectar now known to the hive, followed there by the first of the bees she has recruited.

On coming home with full loads, these in their turn may recruit more bees.

This is what happens when a bee discovers a rich source of honey or pollen not far from the hive, in fact, no farther than 100 metres.

The behaviour of bees discovering nectar or pollen at a distance of 150 metres or more from the hive is very interesting.

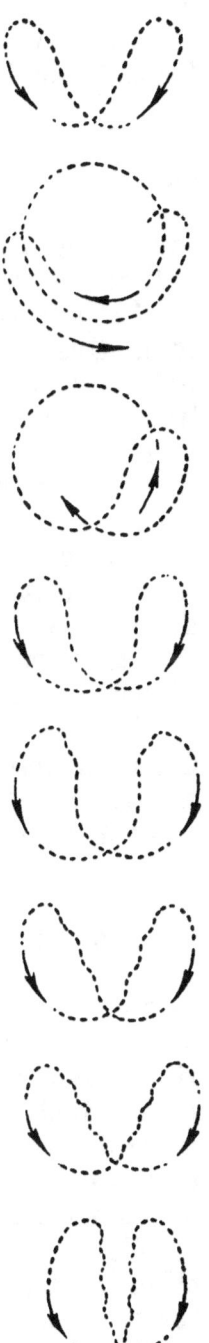

If one follows the movements of a dancing forager, one will see that she almost always runs in curved lines closed in various ways. The diameter of the curved lines is different in different dances but in similar dances it usually is the same. The figures represent variations of the sickle, round, wagging (figure eight) dances and of transition forms between these types

They, too, run into the hive and have their load of nectar taken from them by the home-bees, after which they begin their dance. But this time the dance differs markedly from the one we have just described.

When the source is close by, the bee describes little circles like the letter "o" with a radius not more than one cell; with the source at some distance the bee describes something like the figure eight, with the radius of each half-circle as long as two or three cells.

In performing this complex dance (the first researcher to analyse this dance described it as a half circle to the left, a straight line, a half circle to the right, a straight line and again a half-circle to the left and so on) the bee wags her abdomen during one of the straight runs, and this earned the dance the name of "wag tail" in distinction from the first dance called "the round dance."

For some time the wag tail dance was considered a communication about sources of pollen while the round dance was connected with sources of nectar. We know now that this was wrong, for the two dances may indicate sources of both nectar and pollen.

It is worth mentioning that different races of bees dance differently: today we know of a "sickle" dance, a variation of the figure-eight dance, which, however, has not yet been studied in detail.

Bees bringing abundant loads to the hive dance on the combs. These dances, a peculiar form of reflecting the environmental conditions, can be observed in the hive every day. But can we interpret their objective significance? It is easy enough to ascribe a definite meaning to some movement of the feeler or to a turn of the body, but it is much more difficult to verify if this is not a trick of the imagination on the part of an over-confident observer who thinks he understands nature.

Nevertheless, thanks to remarkable achievements in other branches of biology, it has become possible to decipher the "language" of the various movements in the bee dance.

Long before the interpretation of the bees' mute language was started, I. P. Pavlov discovered an infallible method for studying the behaviour and motor reactions of animals. This method, a great triumph of materialist natural science, enables the research worker objectively to analyse all the higher manifestations of animal life and the entire behaviour of animals. The research worker compares the stimuli experienced by an animal with its reactions to these stimuli and then establishes the laws governing the correlations he has observed.

In setting forth at a session of the Academy of Sciences held September 14, 1921, the results of his long studies of the salivary glands of the dog I. P. Pavlov said that all reflexes or instincts, which are definite and regular reactions of a living organism to definite external agents, are based on the signalization principle.

The history of the discovery and interpretation of such a signal system in bee life is of much interest. We shall speak here of certain temporary connections established between the colony and the external world where the colony finds all it needs for its growth and development. These connections are of the type which I. P. Pavlov considered *organs* for adapting the organism to the conditions of its existence. The story of the study of bee dances, which are just such organs, will at the same time be a story of the discovery of the first links in the "wireless" nervous system of the bee-colony.

A feeder of sweet syrup was placed at a distance of ten metres from an observation hive in an experimental apiary. A piece of cloth scented with lavender was put under the feeder so that the bees associated the location of food with lavender scent. While ten bees brought from the hive were sucking syrup they were marked in a colour code. Through the glass wall the observers saw the bees dance on bringing home the load of syrup.

The recruited bees that came to the feeder were put away in a cage (we now know why) and only the ten marked bees could freely come and go.

Forty-five minutes later the feeder was taken away and two pieces of lavender-scented cloth hid in the grass, one near the hive but a little distance away from the place where the feeder had been and another 150 metres away in the opposite direction.

Four minutes later bees recruited by the ten scouts visited the first lavender-scented piece and during the following three quarters of an hour 340 bees were observed there, while the first bees to visit the second piece came there only ten minutes after it had been placed and during the same period of time only eight bees were observed there.

The experiment was repeated several times with invariable results: the lures placed nearer the hive were discovered more quickly and easily. But perhaps the bees found them only because they were near the hive? To settle this doubt, the experiments were radically changed.

A feeder with bees sucking syrup was placed on a scented cloth 300 metres away from the hive. Eleven marked bees established regular cruises between the hive and the feeder. Then the feeder was removed and two scented pieces of cloth hid in the grass, one at a distance of 300 metres but somewhat away from the spot where the feeder had been, and the other not far from the hive. This time the near-by spot attracted less than a score of bees while the one at a greater distance was visited by more than sixty. The only possible conclusion was that the location was signalized by the scouts. What, then, is the nature of such a signal? This could be determined only by studying what went on in the hive.

Two groups of bees from one colony were marked with two different colours on two feeders. The bees found on the feeder 10 metres from the hive were marked blue, and those on the feeder almost 300 metres from the hive, red.

The observers were posted on two sides of a one-comb observatory hive and awaited the results. They had little enough hope that they would be able to see with the naked eye any difference in the behaviour of the "blue" and "red" bees. Still, before inventing new methods in case no difference was observed, they deemed it worth while to see the results of the experiment. And the results were encouraging. The phenomenon here was brought into being out of its conditions.

Two "blue" bees came to the hive first and started turning round on the comb describing simple small circles.

They were followed by "red" bees which, freed from their loads by home-bees, began describing a figure eight. Many people saw all this afterwards hundreds of times and no doubts remain as to the correctness of the conclusion. The concentration of the syrup produced no effect on the figure of the dances, it was simple circles for near sources and the wagging eights, for the ones at a greater distance. When syrup in the feeder was replaced by pollen the results were the same: the "blue" bees circled and the "red" ones bringing basketfuls of pollen from afar wagged their abdomens and described a figure eight.

In a subsequent series of experiments the "blue" feeder was gradually moved farther and the "red" brought closer, and each new position of the feeders called forth changes in the figure of the dances and in the movements of the marked bees. Thus the dance of the "blue" bees became gradually an eight with smooth movements in the circles and waggings in the straight runs. The dance of the "reds" became more and more like simple circling, and after the feeders changed places the dances were completely changed, the "blue" bees wagging and describing eights and the "red" bees circling and describing "o's."

From these observations, however, it was not clear how the process which I. P. Pavlov described as "changing from the reception wire to the transmission wire" developed.

All that could be seen was that bees excited by the circlings and waggings of the dancer ran skipping after her and repeated her movements stretching their feelers as if stroking the dancer with them. But nothing indicated to the observer how the bees read the communications made in the mute "language" of the dance. Although much remains obscure to this day, it is certain that the dance is a signal, a communication conveying numerous and minute particulars. The rhythm, the number of turns, the speed of the run during the dance—all these have a definite meaning and, we may safely say, can be definitely interpreted.

The timing of the dances has shown that when the source of food is at a distance of 100 metres the dancer makes eleven semi-circular movements during a quarter of a minute; when the distance is 150 metres the number of such movements is nine; with a distance of 200 metres eight, at 300 metres seven and a half, and so on. The greater the distance between the hive and the source of nectar or pollen, the slower the pace of the dance on the combs. When the distance is one kilometre the number of turn is four and a half; for one and a half kilometres four, and for three kilometres two. In a word, the number of semi-circular runs per time unit decreases as the distance increases.

At the same time, the longer the trip the quicker the waggings of the dancer and the greater their number. When the dancer recruits bees for a one-hundred-metre trip she wags her abdomen two or three times a run; with the distance of 200 metres there are four waggings, for 300 metres five or six and for 700 metres ten or eleven. Thus, by observing a bee dance one can tell from what distance she has brought her load of nectar or pollen.

But if the information were limited to indicating the distance alone the recruited bees on leaving the hive, would have to fly in all directions in search of the right spot, and in such a case only very few bees would be able to reach it.

Here the research workers were confronted with events which showed by what infinitely varied ways the development from the simple, the elementary, to the more complex, proceeds in nature. Only recently the mutual anatomic adaptability, the correlation in the body structure of the insects and flowers they forage upon, was considered the most convincing illustration of harmony perfected in the course of centuries through the laws of natural selection. The dances of bees present still more vivid, still more startling instances, and still more subtle adaptations.

Here is how they became known.

Several marked bees were released from a feeder of syrup placed on a mint-scented piece of cloth at a distance of 150 metres from the hive.

As soon as new bees came to the feeder, it was removed and four pieces soaked in mint but not sweetened were put in various directions and at various distances from the hive. Four observers noted all that happened at the four pieces during the following hour. During this time twenty bees came to the piece nearest the hive, at a distance of 15 metres from it; a piece put 150 metres from the hive in another direction was visited by only one bee. Ten bees came to the piece farthest from the hive—at a distance of 300 metres, and the piece put nearest to the spot where the feeder was (90 metres from it) was visited by the greatest number of bees, 38 in all.

This experiment is entered as No. 9 in the experimenters' record journal.

In another and more complicated experiment—No. 10—the feeder was placed 150 metres westwards from the hive (the distance and the direction should be well borne in mind). As was the case in the preceding experiment, the syrup was not scented, but a piece of mint-soaked flannel was put under it.

The marked bees that came to the feeder sucked the syrup, returned to the hive and performed an excited dance there, after which new bees flew to the feeder. While the new, unmarked, bees were sucking the syrup, they were lifted with pincers by the wing and put into a cage, so that only the marked bees could return to the hive.

Some time later the feeder was removed and several pieces of flannel smelling of mint were put at various distances and iu various directions from the hive.

The observers on duty then counted the bees that came to the fragrant lures within an hour.

If the foragers could tell their sisters somehow where they had found the food, then the fragrant piece westward from the hive was bound to attract more bees than those placed in other directions. And such indeed was the case.

Over 80 bees came to the piece in the vicinity of the hive; not a single bee visited the second piece placed 200 metres eastward from the hive, and only one came to the piece 150 metres to the south-east. The piece lying 150 metres to the south-west attracted 41 bees, while the one placed in the same direction as the original feeder, i.e., to the west, was visited by 132 bees, although the piece had been moved 100 metres farther from the hive.

It became clear that the foragers looked for food not just anywhere but in the direction where the feeder had been, and the remarkable fact was that those were not

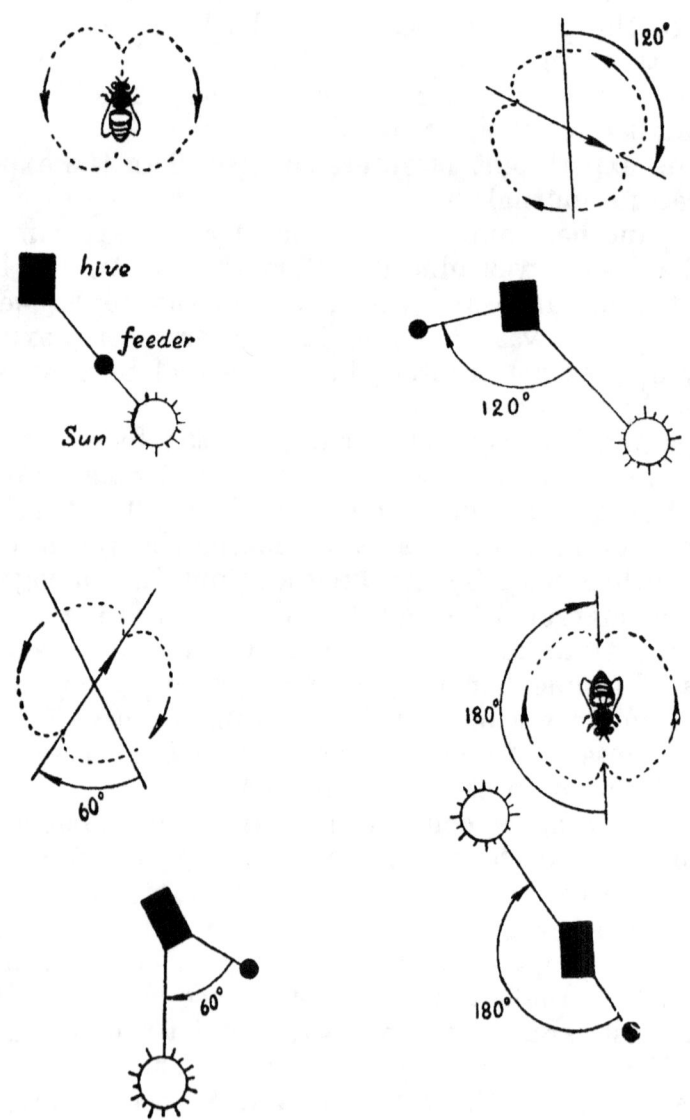

These figures show the patterns described by four dancing bees. The bees came to the hive from the four different points

marked bees but bees that came there for the first time and quite alone.

This experiment was repeated several times with all possible variations, and every time the observations and the counting of the bees on the different pieces most positively confirmed that the bees were looking for food near the spot where *other* foragers of the same colony had obtained food before. This means that recruited foragers leave the hive knowing the direction of flight. Additional investigations have shown that, contrary to popular belief, there is no fragrant track between the hive and the spot where nectar or pollen are found, and bees are not guided by olfactory landmarks in choosing their way.

Today it has been proved that the scouts indicate the direction of flight also through the figures of the dances.

Three points—the position of the sun, the location of the hive, and the spot where food can be obtained—form the apexes of a triangle in the air, two points—the location of the hive and that of the food source—being constant and the third point variable. The angle formed by two straight lines—the one from the hive to the food source and the other from the hive to the sun—is the key to the signal. The magnitude of this angle—the solar angle—determines the manner in which the straight runs in the "figure-eight" or "sickle" dance are performed.

Students of "bee language" remarked long ago that bees performed the wagging "figure-eight dance" differently at different times. The figure eight made up of two "o's" can be described in different ways: bees may perform the runs between the semi-circles with their heads up, in which case the right semi-circle is described clock-wise and the left anti-clock-wise; the runs may be performed with the dancers' heads down and then the left semi-circle is described clock-wise and the right anti-clock-wise.

During an experiment (early in the afternoon) the observers noted that all the bees marked blue coming from a feeder to the north-west from the hive danced in the same way, while other bees of the colony performed dances differing one from another and from the dance of the "blue" bees.

The question naturally arose, why bees coming from other places danced differently. This question was the more pertinent as by the evening the bees returning from the feeder performed the same figure-eight dance, but in different ways: the "eight" was described in a different way as compared with the day-time. The dances of other bees were different from their own dances in the day-time and from those of the bees returning from the feeder.

The next day two feeders were placed: one to the north-east where the bees were marked blue and the other to the south-west with red-marked bees. What happened was that the "blue" bees performed the straight run with their heads down, the "red" bees with their heads up.

For further study the research workers engraved on the glass walls of the observatory hive a network of horizontal lines, vertical lines and lines at different angles to them. This enabled them to establish more or less precisely the angles made by the straight runs connecting the semi-circles in the long-distance "figure-eight dance."

In the experiment that followed, bees coming from four feeders situated at the four cardinal points were marked in four different colours. Observations started at noon. At that time bees from the southern feeder performed their dance on both sides of the comb with their heads up; bees from the northern feeder held their heads down, bees from the eastern feeder bent their heads to the left and those from the western to the right. The positions of the dance changed during the day in accordance with the change of the solar angle.

All the changes were so obvious that it was possible mathematically to establish the figures of the dance for each hour of the day in connection with the location of the feeders On the combs under the engraved glass the bees described figures which formed a regular solar azimuth for the foragers.

The recruited bees automatically repeat in the dance this trigonometrically presented address and follow it in flight, and that is why they can fly to the source of food guided by the sun, without anyone to lead them.

It should be added that not every bee bringing a load to the hive starts dancing: a forager dances only when

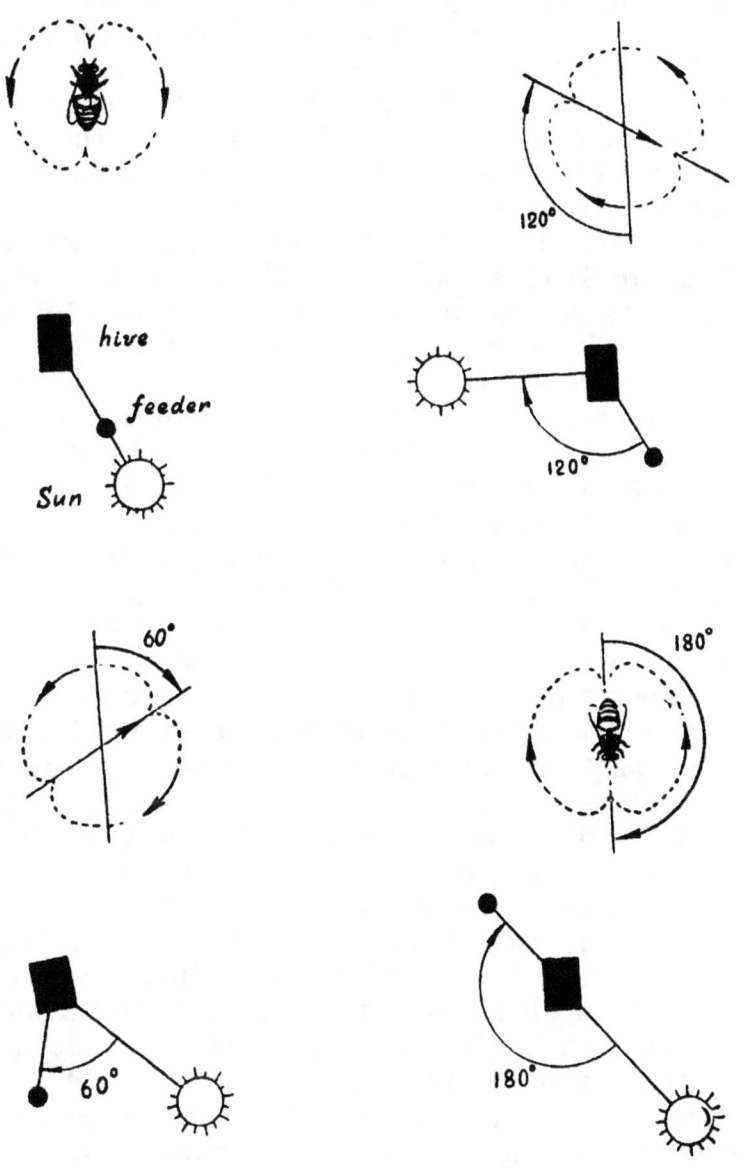

Figures described by a dancing bee coming from one place at different hours of the day

the source of nectar or pollen is sufficiently abundant, and the more abundant the source of food the longer she dances, recruiting the greater number of bees.

Thus, if a bee is put on a piece of blotting-paper, sparingly sprinkled from below with syrup, so that it can be sucked with great difficulty, she will not dance and recruit other bees to this spot. On arriving at the hive and passing over her load to the home-bees, she herself may, however, return to the blotting-paper.

Additional information obtained through experiments completed in 1948 shows that with a contrary wind the bees dance as though the source of food were situated farther from the hive, while with a fair wind the dance indicates a place nearer the hive than really is the case.

New experiments conducted in 1950 in a mountainous area showed that if on leaving the hive the forager had to fly upwards the dance was performed somewhat more slowly, as if the needed spot were situated farther away and, on the contrary, if the road lay downwards the dance was quickened as if indicating a shorter distance.

The most startling, however, were the results of experiments completed in 1952, which showed that the behaviour of dancing foragers depends on the state of the colony's food stores. When there is little honey and beebread in the cells, the bees dance energetically calling their sisters even to meagre sources of food and when there is enough food in the combs, poor sources are no longer announced.

At the same time, other observations made at the apiary in Gorki Leninskiye show that bees can be made to dance by complex, "partial" irritants. After all forage trips had been over (on October 2, 1949) the quilted cover was removed from an observatory hive and a powerful bulb was brought near its glass wall, which made about a dozen bees in the centre of the comb perform a short but energetic dance. The experiment was subsequently repeated several times and each time the abrupt change from darkness to a bright light induced a certain number of bees "slumbering" in the cluster to dance, thus revealing the conditioned nature of this reflex: it developed through associating foraging with light.

Further we shall describe the dance by means of which

scouts sent out before swarming report on the whereabouts of the new home they have found for the swarm.

The curious thing about all this is that no dislocation of the hive, not even changing the position of the comb from vertical to horizontal could prevent the bees from accurately solving the problem before them: with every change in the position of the hive or the combs the direction of the dance was changed accordingly and correctly. Only dancing upside down on the lower plane of a horizontally-placed comb was too much for the bees and they lost their bearings.

Now we shall sum up the results of all the experiments.

When a food source is near the hive, the forager has no time to commit to memory its location and the signal merely states that food can be obtained. The communication is made through a circular dance, which, translated into human speech means approximately this: "There is food close by, go look for it near the hive and you'll find it easily. Don't waste the time at home—the flowers are full of nectar."

The forager indicates the direction of flight when the spot is at more than 100 metres from the hive. The direction the bees learn from the wagging dance, whose rhythm and steps vary according to the conditions mean something like: "There's food! Mind, you'll have some distance to fly for it. Repeat my movements and mark the tempo, the direction of the semi-circles and of the straight runs. Take your bearings and away with you while the sun has not changed its position and mixed everything up. Go. Meantime I'll call other bees—there's lots of excellent food there!"

All this makes bees seem very clever, although not more so than a dog suffering from worms and instinctively eating Artemisa vulgaris, a vermifuge which it finds among a multitude of other herbs and grasses. At the same time we must admit that the field activities of the honey-bee offer an example of an extremely fine instinct and of exceedingly complex and clearly perceptible conditioned connections in time.

Professor Karl von Frisch tried to use the above laws of bee behaviour to substantiate a frankly idealistic conclusion that there exists in living nature a super-wise "biological reason." As we know, he was not the first to proclaim the superiority of natural instincts over consciousness. His conclusion is worth noting because V. I. Lenin mentioned a similar case in his famous *Materialism and Empirio-Criticism.* Saying that "one school of natural scientists in one branch of natural science has slid into a reactionary philosophy, being unable to rise directly and at once from metaphysical materialism to dialectical materialism," Lenin observed that sometimes science develops towards "the only true method and the only true philosophy of natural science not directly, but by zigzags, not consciously but instinctively, not clearly perceiving its 'final goal,' but drawing closer to it gropingly, hesitatingly, and sometimes even with its back turned to it."

The new facts of bee biology established thanks to a consistent application of I. P. Pavlov's method of the study of conditioned reflexes, which has enabled us to interpret bee "language," bear witness to the potency of this teaching as a means of knowing nature.

It is just this objective method successfully applied in the further study of bee behaviour that has explained how, along with the "dance language," bees use the "language of flowers."

FRAGRANT BEACONS

How Bees Find Their Bearings on Flowers. An Examination in Geometry. The Fragrant Codes and the Transmission Thereof. What Non-Fragrant Flowers Smell Like. The Complete Signal Code.

Now suppose that raspberry has started blossoming in a glade. Raspberry blossoms are homely and inconspicuous; the bushes are surrounded with a wilderness of flaming-yellow butter-cups, dandelions, purple-red catchflies whose blossoming time is drawing to a close, blackbloods just bursting into blossom, pink gymnadenias and sonchuses, sky-blue bluebells, and snow-white mountain clover.

Why is not the bee dazzled by all this wealth? Why does she fly indifferently over the lush live carpet, every patch of which cries out to her in the language of bright colours and heady perfumes? Why does she calmly alight on the raspberry blossoms which are so insignificant as scarcely to be worth the name of flowers? It would be too much to suppose that the dancing fielder has told her sisters, in addition to the whereabouts of the source of food, the nature of flowers it is to be found in. We simply cannot suppose the language of bees, however rich, to contain definitions for the multitude of flowers existing in nature. For all this, the recruited bees unhesitatingly choose from among the mass of flowers in the glade the homely raspberry, or the catchfly, or the bluebell, although this last is no great nectar-bearer.

It is well known that in an ocean of yellow flowers in the glade a bee will quickly find the yellow flowers of sonchus she is interested in. The mistakes we shall mention here are of a nature that confirm the general rule.

Describing in his *And Quiet Flows the Don* the first meeting between Listnitsky and Bunchuk, Mikhail Sholokhov says that as Listnitsky stopped near the birch-trees "a bee settled on the brass hilt of his sabre." Here the bee was taken in by the gleaming yellow of polished brass.

The mistake of the bee serves here as an additional feature in the description of the autumnal background of the scene: "the grasses were becoming pink, their vivid autumnal colours speaking loudly of coming death." And indeed, in summer with an abundant honey-flow on, field-bees, as a rule, make no such mistakes. More than that: if the dancer comes from an enamel, china or glass feeder containing syrup, the recruited bees will find it in the thick of blooming grasses and will alight not on the flowers but on the feeder, although a feeder is not like any flower in the world and the syrup does not smell of any flower.

There is no doubt—it has been established experimentally—that the colour and scent of the flower which a bee has worked are of importance for her subsequent visits. Special experiments were carried out in which bees were trained to take syrup with a strong smell of jasmine from a feeder placed in a blue box. Then the box was moved some distance from the feeder which was put into a yellow box. In this way the attraction "blue jasmine" was divided into two elements and the bees were given an opportunity of showing their preference for either the colour or the smell. Marked bees coming for a new helping of the syrup flew straight to the empty blue box. However on coming closer, they veered away without entering the empty box and, after several orientation circlings, turned in the direction of the unfamiliar yellow box with the familiar smell of jasmine.

The behaviour of bees in this and other experiments, both with artificial flowers and with natural flowers with their petals removed, has shown that from a distance bees determine their bearings by a familiar colour, but at close range by a familiar smell.

Incidentally, similar experiments were carried on with bees deprived of their feelers, and such bees entered the empty blue box and looked for the feeder there.

The question whether the size and shape of the flower from which bees take nectar is of any significance remained long unsettled. The experiments with artificial paper and cloth flowers yielded no definite answer to the question, and then a series of tests was started, jokingly called "a geometry examination."

A feeder with syrup was put on a blue circle placed on a white table and bees were trained to go to it. After a sufficient number of bees had for some time been flying regularly to the spot, the feeder was removed and two similar but empty feeders put side by side on the blue circle and a blue triangle. Observers were posted to see which figure would attract the bees. The researchers were amazed at the length of time a bee would hesitate before alighting on one of the figures. Similar experiments were carried out to study the ability of bees to distinguish other geometric figures. Finally it was established that although purely geometric figures are not met with in living nature and are new to bees, foragers can be trained to distinguish between vertical, sloping and horizontal stripes, between triangles and polygons, between equilateral and scalene triangles, etc. They distinguished differently-coloured triangles and triangles of different sizes fairly well. In an experiment in which one of the two identical figures was stationary and the other was often displaced, the bees were taught to distinguish between them, although on coming to the spot from different directions they saw the two figures differently each time.

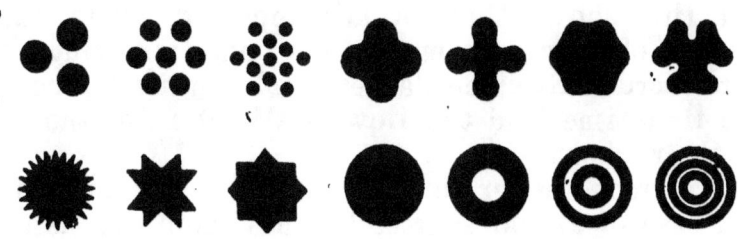

The various figures that were used in the "geometry examination." The area and the length of the sides of each figure were measured, the latter being found to influence most the behaviour of the bees

For a better study of bees' behaviour on the spot of the take and of their ability to find the source, bees were trained to suck scentless syrup, little drops of which were put on a tiny piece of very thin glass placed on a big sheet of glass so that it was all but indistinguishable. It would seem that the smooth transparent surface of the glass presented no visible marks for bees to orientate themselves by. But the foragers found the drop of syrup and sucked it up. When the bee returned from the hive she found another drop on that very spot and again sucked the inexhaustible drop. Then the research workers wanted to see what the bee would do if the drop was not on the little piece of glass but under it. Would she find it there? The forager did find the drop under the bit of glass, and since the glass was very light, she managed to lift it by putting her proboscis under it and to suck the nectar as before. Then the drop was put under the glass but far from its edge, so that the bee could not reach it with her proboscis at once. The bee inserted under the glass first her proboscis, then her head and thorax and contrived to suck the sweet drop.

After the same bee had been observed doing this several times, the drop of syrup was replaced by a drop of water. On her return for more syrup, the bee inserted her proboscis under the glass but recoiled as if she had burnt herself. Then she tasted it once more, as though to make sure, and flew away. She did not visit the spot any more.

All this shows that in her repeated visits the bee makes use of numerous means of orientation. But how do foragers recruited by the dancers and coming to the source for the first time find the flowers? What helps the bees to make their choice?

The correct answer to the question is the more important in case of a near source, because then the dance announces nothing but the existence of one. At a distance of 50 metres when no definite address is indicated the bees have to investigate an area of nearly a hectare. Some signals are absolutely necessary to enable the recruits to find the needed flowers without wasting time and energy on investigating all the flowers they see within such an area. What are those signals? When does the dancer make them? How do the recruited bees perceive them?

Here we must recall one detail in the description of the dance on the comb: "the bees hastily follow the dancer and seem to stroke her with their outstretched antennae, repeating her every movement."

Here lies the key to the riddle.

While the forager was busy in the flower sucking nectar from the deep-lying nectaries or packing her baskets with ripe pollen, the flower imparted its fragrance to her. From the first flower she went to another of the same species, then a third, and so on, and the fragrance clinging to her grew stronger and stronger. (Thus we see that the fixation instinct performs one more function in addition to those mentioned above.) The fragrance of the flowers that had supplied her with nectar and pollen permeated the bee's fuzzy body so thoroughly that the bees stroking her in the hive smelt with six thousand olfactory pores of their antennae the fragrance of the flowers in the field. On leaving for the field the recruited bees knew perfectly well what scent to look for among the multitude of various aromas filling the air.

Usually the immortelle is not visited by bees. But after some marked bees had been fed on syrup smelling of immortelle, bees recruited by their dances discovered this flower among the 700 other plants blooming in the experimental plot.

If the scent of a flower is too weak or if the distance is so great that it becomes lost, the bee can preserve it in the nectar she carries in her honey-crop as if it were a corked bottle of perfume. This holds good of both nectar and pollen: pollen loads also smell of the flowers although not so strongly as nectar.

The bee's sense of smell enables her to discover the required scent among many others; precise experiments have shown that bees discover it even when greatly diluted.

The smell need not be pleasant to man.

At a collective-farm apiary, bees were once fed by sugar smelling of petroleum and the next day the work-shops of the neighbouring machine-and-tractor station and the filling-station were invaded by thousands of bees which crawled over petroleum-soaked rags used for polishing machines, over machine-parts wiped with petroleum, over the men's overalls, over the tanks and barrels containing this liquid. Such is the attraction of the fragrant lures even when the term "fragrant" can scarcely be applied to them!

On learning the direction of flight from the dancer, a bee draws her feelers through the combs on her legs, cleans her eyes and takes off from the landing-board. Obeying her instinct, she flies in the required direction following the sun compass at a speed of up to one kilometre a minute.

She flies over trees and bushes, over fodder and cereal crops from which rises a medley of attractive smells. Several times the bee may sense here the smell of raspberry which is exactly what she needs, but until she has covered the distance indicated by the dancer she will not mind the "language" of scents.

This is an important adaptation preventing the bee from responding to the call of lonely bushes where little nectar can be gathered or of clumps already worked by groups of other bees. Only on covering the specified distance does the bee start looking for her fragrant lure which by its scent will lead her to her destination. Guided by it, the bee flies past all the butter-cups and bluebells, catch-flies and gymnadenias and unfailingly reaches her objective.

The poet Nikitin said:

> *Golden bees hover*
> *Over fragrant flowers. . . .*

But we know now that bees can hover over flowers that have no scent and that in some way or other they can reach scentless flowers or flowers whose scent is scarcely perceptible.

In the chapter on bee dances we mentioned that a forager coming from a poor source does not dance in the hive. Now we shall speak of the interesting discoveries made during the study of bees' behaviour in the case of a scentless take.

Two feeders with a solution of sugar in water were put at equal distances right and left of a hive under observation. The sweet water had no smell perceptible either by man or bee, which was ascertained with the help of special investigations. The right feeder offered a plentiful supply of sugar solution and the bees that visited it were marked white. The left feeder—a piece of blotting-paper moistened with the solution—was a scant source. The bees visiting it were marked blue. The "white" bees danced in the hive while the "blue" bees sucked the solution with difficulty from the blotting-paper, brought it to the hive but did not dance. It seemed probable that bees called out by the "white" bees should either fail to find the feeders at all or, if the mere presence of other bees attracted them, should fly in equal numbers to the plentiful and the poor source. But actually the feeder with syrup was visited by ten times as many bees as the sweetened blotting-paper. And the fact can be easily explained: first, the more bees are concentrated in one spot, the more perceptible are the supersonic signals we have had occasion to mention above, and secondly, if a source naturally has no odour bees themselves can make it odorant.

The anatomy of the honey-bee has been studied for centuries and it would seem that no cell of this insect's body has had a chance to escape the notice of anatomists and histologists. And yet, N. N. Nasonov, in 1883, discovered a new gland—a scarcely perceptible fold near the back tip of the abdomen. When the bee protrudes it, the fold can be well seen and the glands concealed in it exude a smell which to some seems like the scent of the well-known honey-bearing plant, melissa, and to others, that of quince fruit.

The females of different insects are provided with scent-glands for attracting males. There are many descriptions of experiments in which males were attracted by scent-glands cut out of a female's body and took no notice of females crawling about with their glands removed.

But what is the function performed by the glittering scent-gland roll in the life of infertile worker-bees? Of what use are scent-glands to bees? When and how do the bees make use of them?

For a long time no reply could be found to these questions and it was but recently discovered that bee-scent constitutes one more link in the system of signals leading to the source of food.

When flowers are rich in nectar or when a feeder is full of syrup, bees suck the sweet liquid pumping it into their honey-crops with all their might. Their abdomens move in a characteristic way, now rising and now extending, at the same time baring and expanding the white gland, whose secretion permeates the spot of foraging. In this way a "fragrant beacon" is left there.

When the source is poor the bees work less energetically: the gland remains inactive and the bee-scent is not diffused on the spot, which, consequently, does not attract other bees.

Thus we have a more or less clear picture of the signal system: the round dance calls bees out in search of a source near the hive. On the day the raspberry bursts into bloom, the majority of recruited bees will work raspberry bushes, not necessarily those that have been visited by the first foragers, but all bushes within the near range where the bees probably sense the smell brought into the hive by the dancers.

Direction is communicated only for more remote trips, since without knowing the direction bees would have to explore an extensive territory and the percentage of finds would be insignificant. The scent diffused by the foragers serves as an additional reinforcement of the fragrant beacon the recruited bees are looking for. When the source of nectar is exhausted the foragers' glands give off no scent and fewer and fewer bees come to the spot until nectar in the flowers dries up altogether.

THE MAIN HONEY-FLOW

More about Experiments with False Lures. Unexpected Good Luck of Thriftless Foragers. Landmarks Along Bee Track. Fixed Forage Areas of Individual Bees. A Bee "Refusing" the Feeder. A Story about a Loss of Linden Honey. The Forager and the Filament.

The foregoing chapters describe how foragers inform their sisters of the whereabouts of the source. Let us go back to the results of experiments showing that dancing bees indicate by circles and figure-eights the way to food. Take by way of example Experiment No. 10. Here the greatest number of foragers—132—came to the feeder nearest the spot where the initial feeder had been. But during the space of an hour 123 bees—almost as many as in the first instance—came to the four feeders in other places, which means that at least half of the recruited foragers were flying in search of the source at random and their trips yielded no results. This means that the wonder we have just gone into raptures about, is not so wonderful after all. And in this respect Experiment No. 10 is no exception. Under Experiment No. 9 the first piece of cloth placed not far from the initial feeder attracted 38 bees while the other three pieces together attracted 31 bees.

Under Experiment No. 11 in which six false lures were employed (we have not described it in this book) 109 bees flew in the right direction and 112 bees went to the other places.

So after all, the "chart" the foragers receive in the hive does not lead all of them to their goal: every experiment in which more than three false lures were used showed that almost every second bee recruited failed to find the source. We may therefore suppose that under natural conditions with many flower-beds scattered around the hive and not merely three fragrant lures, still fewer bees reach the spot indicated by the dancer.

The simple apparatus for measuring the quantity of liquid food a bee can hold in her honey-stomach. It consists of a transparent graduated capillary pipe with a plastic flower at the end. A bee settles herself on the flower and extracts syrup from its "nectary"

What kind of adaptation is this if its efficacy is so low and if it involves such a waste of energy on useless trips of foragers? Since the main honey-flow, when the staple honey-bearing plants in the given locality are in bloom, is often of a very short duration how could the most diligent bees collect enough honey if only one forage trip out of a series were useful?

But actually such is not the case.

In the first place, not all the bees recruited by the dancer and flying in different and wrong directions merely waste their energy. Some of the foragers scattered all over the neighbourhood come across new sources and extend the pasture area of the colony by new flowering lawns and flower-beds, thereby adding to its sources of food and to some extent making up for the expenditure of energy by bees returning without a load. Then on returning home the bees which have not reached the goal and found no new source soon crowd again round dancing bees trying to learn a new route from their dance and using their antennae with their six thousand olfactory pores to remember the scent of the new source. Once the signal is received, they again clean their antennae and eyes with the brushes on their legs and again mount into the air and fly away guided by the solar compass.

The more lucky among the foragers, those that have found the source at the first attempt and brought home two or three loads, fly to it along the well familiar road and during repeated trips look more often on the ground than to the sky.

Let us relate how this was discovered.

In an experiment a saucer with syrup was placed at some distance from the apiary situated in flat country, and the shortest road to the feeder was marked by conspicuous landmarks. For several days the feeder was put on the same spot and regularly refilled. Bees were extremely active on it and visited it from morning till night. On the sixth day, when nightfall put an end to their flights along the beaten track, all the landmarks were replaced so as to lead in a direction different from the one where the feeder was. Where would the bees fly the next morning?

The bees followed the landmarks and arriving at the last one circled for a long time about it in search of the feeder. Meanwhile the saucer was full and stood in the old place, but it was long before a bee was seen there. The meaning of all this is clear: after a route to the source has been traced, the foragers flying to a familiar spot orientate themselves by salient features of the terrain, the location of the source itself being the last landmark on the route.

An interesting observation was made about twenty years ago at the honey-bearing plant plot of the Experimental Apiary in Tula. Sainfoin was sown on several beds separated by intervals of 50 centimetres. The plants grew and interlaced, forming one unbroken bed. When the blossoming time came the observers marked the bees that worked the flowers. The bees found on the first bed were marked white, on the second red, and on the third yellow. The foragers filled their honey-crops with nectar and flew away to return some time after; but the curious fact about them was that the bees marked white came to the first bed, the bearers of red marks to the second and yellow to the third. This might seem very improbable and yet the bees kept to their respective flower-beds not for hours but for several days on end. When, however, few blossoming plants were left, the unseen boundaries were broken and the bees began collecting nectar on all the beds indiscriminately.

The experiment was repeated on beds of other flowering plants, in particular with blue weed, and the results were similar.

More than that: while blue weed was still in bloom, lindens started blossoming, but the marked bees continued to work the blue weed even more vigorously than before, as if infected with the excitement of the bees collecting abundant loads of nectar from the lindens.

Since these important facts were noted bees' fixations to natural sources of food have been repeatedly studied under various conditions.

The conclusions arrived at as a result of the experiments on different beds at Tula were later confirmed on a larger scale by experiments conducted by Sardar Singh, a young Indian scientist. He kept observations on a field of buckwheat of about 16 hectares, and a lawn overgrown with pink clover, dandelions and other plants, and an old apple-orchard. Observers posted in the field, in the lawn and in the orchard marked on plans the exact spots the bees visited, their roads from one flower to another and the spot from which they flew back to the hive. This was a hard task, as was the summing-up at the end of each day of all the records, when the work done by each bee had to be reconstructed piece by piece. In spite of all difficulties, the work was continued throughout a summer season with occasional short breaks.

It was these observations that helped finally to establish the fact that each individual forager has a fixed, more or less limited working area in the field, meadow or orchard. Bees worked areas differing in size but each had her own "sector," or, to put it differently, there was a bee for each plant. In the clover field the average size of bee-

areas was 12 sq. metres, in the buckwheat field—8 sq. metres; one bee worked about 5 sq. metres of goldenrod and 18 sq. metres of lotus corniculatus.

When little nectar was left in the flowers or it became less sweet, the forage area expanded quickly, but fixations for definite sources could still be observed.

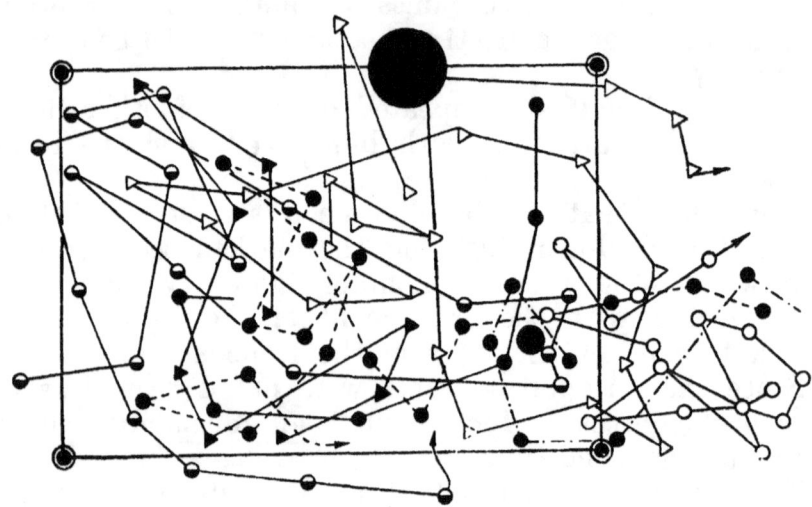

A forager was marked on a dandelion flower at noon, May 26. This spot is represented by a small black circle (the big circle represents a tree). In the course of four days (May 27, 28, 29 and 30) the marked bee was discovered several times in that area. The bee's journeys are represented with different signs. The observations noted here were made by Sardar Singh, an Indian scholar, and prove that foragers have their fixed areas of bee pasture

If a separate chart is made for each forager and the spots where she alighted on flowers during the period of observation are marked in red while the points from which she took off at the end of each trip are marked in blue, a very interesting picture will be the result: the red marks are concentrated in one spot, sometimes coinciding, while the blue marks are more scattered. Dozens of such schematic charts show that everywhere the points of arrival are more concentrated than the points of departure.

Within her forage area a field-bee usually works, as already mentioned, the flowers of one particular species, taking no notice of blossoming plants of other species.

A bee may visit the flowers of only one tree during her entire field life, as in orchards with big fruit-trees whose crowns are several metres in diameter. In a buckwheat field, too, bees have been observed to work their own areas exclusively. A bee carried by the wind away from her particular forage area to another patch of the same buckwheat did not work the flowers there. She flew against the wind towards her "own" plants and started collecting nectar "on her own grounds" as it were. All these important details were very carefully observed and noted, but one more proof was needed for a final conclusion.

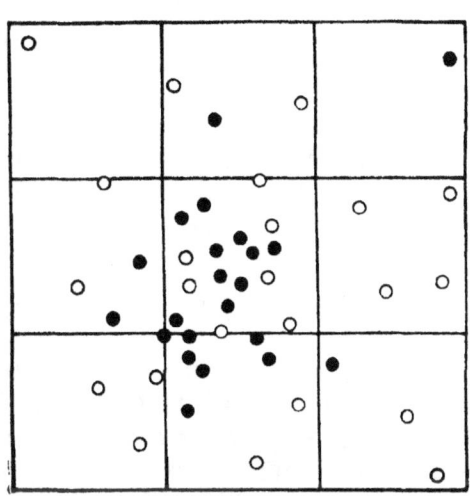

Dark circles show the spots where the bee alighted and light ones, spots from which the loaded bee took off for home. (Observations made by S. Singh, May 9-18, 1953)

Over a hundred little tables with feeders containing sugar syrup were placed in staggered rows over a large lawn near an apiary in Rothamsted. Thus the lawn near the apiary was turned into a kind of meadow, an extensive source of honey.

Observers posted at each table marked each coming bee with their particular colours. Soon it became quite clear that each bee visited only one feeder and found it unfailingly among a number of other feeders in no way differing from hers. If a bee made a mistake now and then, she alighted on the feeder next to her own.

But things went on in this way only as long as there was enough syrup in all the feeders. No sooner had an empty feeder been put on a table than the bees abandoned it and began flying to neighbouring feeders, exactly like the foragers in the experiment with onobrychis where, too, the boundaries were forgotten as the blossoming time came to an end.

When, however, one of the feeders was filled with more concentrated syrup the bees did not abandon their accustomed feeders for this one, except when a forager came here by mistake. In such cases, the bee filled her honey-crop with the sweeter syrup and made several orientation flights over the table, after which she always came back to that feeder and forgot her old source. The others visited their accustomed places.

To give a clear idea of the practical significance of the above peculiarities in the behaviour of field-bees, we shall quote an instructive episode from practical bee-keeping.

A plot was sown to phacelia near an apiary of a state farm in Lunino District, Penza Region. The plants started blooming in the second half of June and, since no other sources were available at that time, the bees from the apiary were buzzing over the spot from morning till night. But the control hives on scales showed little increase: not more than 200-300 grammes a colony a day.

Then at the end of the month a wealth of blossoms appeared in a linden-grove, also near the apiary. The main honey-flow came. But the daily increase of honey was as poor as before, because the bees continued visiting the phacelia instead of flying to the lindens.

Three days passed by in this way and the apiarists became alarmed at losing the linden take. So they had the phacelia mowed down and expected the bees to start working the lime-trees. But nothing of the kind happened!

Disappointed bees were hovering from early morning over the mown and faded phacelia and furiously attacked anyone who dared approach them there or at the apiary. The foragers continued flying to the barren phacelia patch for many days afterwards although close by the linden blossoms were full of nectar and diffused their heady and inviting perfume.

To sum up: all the facts quoted above clearly show that after a bee on her first trip afield has alighted on a flower (or a feeder) and filled her honey-crop for the first time in her life with nectar, the spot where she has found food for her colony becomes specially attractive to her.

This spot lures the forager, and to it she steers her course on leaving the hive. Upon bringing home her load, the bee may gyrate and describe on the comb the figure eight or the sickle of the "recruiting dance," but she herself will pay no attention to the dances of other bees. Such a bee does not fly *in search of* a source, but, guided by marks on the ground, plies between her source and the hive like a shuttle, and pumps the nectar from the flowers into the combs until her source is exhausted.

This is exactly what is usually observed during the main honey-flow, when one can hear from afar the incessant buzzing of numberless fielders hurrying in a bee line from the hives to the flowers and back.

All the above leads us to the conclusion that the meaning of the dances is limited and that they are not the one and only adaptation controlling the field activities of honey-bees.

The colony owes to the foragers' dances only the first drops of honey brought during the first trips; and it is to the fixation instinct keeping each forager at her own area that the colony may owe as much as a spoonful of the precious liquid.

A forager's dance introduces a bee that is .or field life to a new duty, thereby introducing her ı ꞁ ꞁ the last phase of her individual development. All the circles, sickles and figure eights only excite new bees to flight and lead them to sources of food. Repeated visits to the same spot are regulated by other instincts, which enables us to draw an analogy between a bee and a filament of a plant by means of which it is attached to the soil whence it derives its food.

The similarity between the function of plant filaments fixed in the ground and the routes of foragers, fixed in space, is not merely on the surface.

In his *Life of Plants*, K. A. Timiryazev dwells in detail on the great physiological importance for plants of the growth of the roots in length. Thanks to this peculiarity, a root is able with the utmost economy of the building material to "come into close contact with the greatest possible number of soil particles," as Timiryazev puts it.

It has been calculated that the filaments of a living wheat plant make up a surface exceeding a hundredfold the area of field occupied by one wheat plant. At the same time all these filaments, about 20 kilometres long, can be put in a thimble, their volume being about $1^1/_2$ cubic centimetres.

As to the bee-colony, with the help of its three or four handfuls of winged foragers it can cover an area several million times exceeding that of the hive. It is thanks to this that the bee-colony can make sufficiently large stores of a food dispersed about in myriads of microscopic droplets secreted by those ephemeral beings—plant blossoms.

FROM THE FLOWER TO THE HIVE

THE ROAD HOME

How Flying Insects Find Their Bearings. Savelovsky Railway Station and Pushkin Square: Two Starting Points. Bees That Have Lost Their Way. The End of the Mystical "Homing Sense" Legend. Is There a Feeling of Animosity between Bees from Different Colonies? Why Field-Bees Are Attached to Their Homes.

Each fielder sees around her other bees flying with loads, buzzing in all directions, and at the same time emitting supersonic signals inaudible to man. When a bee has filled her honey-crop with nectar and her pollen-baskets with pollen, she feels an urge to return to the hive.

A bee that has travelled five kilometres from her hive has put between herself and her home a distance about half a million times her own size, but for all that, this living particle of the colony that has flown so far away from it in search of food, this minute grain of sand lost in the green ocean of plants, confidently starts on her way home.

Bees acquire the ability to find their hive in orientation- and play-flights. If several young bees that have left the hive for the first time are taken some 150-200 paces away from it they will not find their way home. When Academician N. M. Kulagin (at the time he was Professor of the Timiryazev Agricultural Academy in Moscow) marked several old foragers with red, brought them in a cage about 4 kilometres from the apiary—to Savelovsky Railway Station—and set them free there, the bees, after making several orientation circles disappeared from view and five minutes later were seen landing on the porch of their hive.

A few days later, a group of other old bees was taken in a cage to Pushkin Square—a distance of five or six kilometres from the apiary. On leaving the cage the bees were seen circling and then lost sight of. But they did not find the apiary. Failing to return home, the bees returned to their starting-point—the cage —spent the night in it and in the morning made another unsuccessful attempt at finding their hive.

A bee working her wings at the entrance with her legs free of pollen loads is a fanning bee

The bees' "sixth sense," their orientation ability, was studied for a long time and in various ways. The research was aimed merely at establishing the greatest distance from which bees can find their home and did not touch upon the nature of this ability.

The firm belief was held for a long time that bees are endowed with a mysterious, not to say mystical, "homing sense" which directs them in their flights and brings them safely back to the hive.

But N. M. Kulagin's experiments can be quite realistically explained if we admit that before the experiment the bees might have visited the gardens in the vicinity of Savelovsky Railway Station and not gone so far as Pushkin Square. This is exactly where the key to all these observations lies.

Numerous experiments carried out recently show that a bee coming home from her first trip follows the road she went by. If, however, a perfectly new road brings her to a spot she has visited before, she chooses for going home the shortest and most convenient road of all known to her.

This is why a bee carried away from the hive but a short distance will find her way back only if in her search for a way home she finds herself in a spot well known to her from previous trips. Even old field-bees cannot find home from a place they have never visited before. Since in this case, too, they can find their way home only from an already known spot, they are found in the spot they started from. This explains why N. M. Kulagin's bees set free from the cage in Pushkin Square went back there on failing to find their way home.

But then, to return home, it is not enough to find the locality of the hive. A modern apiary consists of dozens and hundreds of standard hives which are as like one another as twins. A bee must be able to recognize her own hive. True, hives are often painted different colours (of course such as bees can distinguish). But what about bees living in hollow trees in a forest? They find their home among thousands of trees.

A bee with pollen loads on her legs fanning her wings on the landing-board is a forager drying moist pollen

A thorough study has been made of this ability of bees.

When a forager comes home from a short-distance trip an observer can easily detect the presence of the "sense of direction" in the bee.

If moved aside—to the right or to the left—together with the feeder, bees that have several times visited a feeder of syrup placed in an open space not more than a hundred metres from the hive will make for home along a line parallel to the road they had previously taken to the feeder. This road will not bring them home.

This instinct manifests itself still more convincingly if a feeder that stood a hundred metres in front of the hive is removed with the bees on it to a hundred metres behind it. The bees will fly home in the direction that would be right if the feeder were in its old place, i. e., actually they will fly away from the hive.

After what has been said, nobody will be surprised at learning that when a feeder standing a hundred metres in front of the hive was put (with bees on it) on the cover of their own hive, the bees made a bee line for "home" which they expected to find a hundred metres away.

Other aspects of the flying and orientation instinct of home-coming bees, too, have been closely studied. Among these, bees' ability to distinguish colours has formed the object of detailed investigation. In recent years many experiments have been conducted with white hives, hives of different colours, and hives whose front walls were cov-

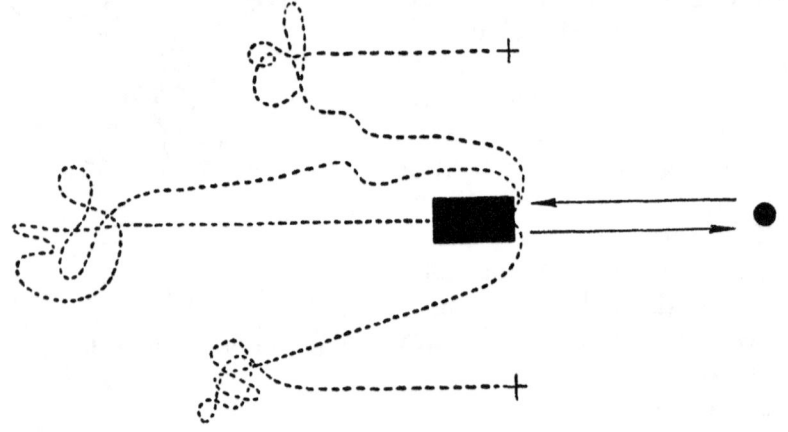

Diagram to the experiments described on page 225. Bees came from the hive (rectangle) to the spot designated by a circle. Crosses designate spots where the feeders with bees on them were replaced, and the dotted lines show the route the bees took in their search for the hive.
We must remind the reader that the experiments were made in a plain devoid of salient landmarks

ered with a piece of plywood painted in two different colours and the landing-boards also painted in two colours, for instance, blue and yellow. This last device enabled the research workers to preserve the familiar smell and position of the hive and at the same time to change its outer aspect by changing the respective positions of the two colours.

These and other experiments—such as changing the colour of hives surrounding the one under experiment, changing the positions of the hives, removing the whole series of hives to a new place while preserving their respective positions, or changing both the place and the positions—by making the bees reveal ever new aspects of this instinct, helped the research workers to arrive at definite conclusions.

Here is what they have learnt:

As in her journey afield, a home-coming bee makes use of the solar compass and landmarks on the ground which lead her to the source. The bee's compound eyes, in which the light-absorbing surface of the walls extinguishes rays coming at an angle and which perceive only direct rays, are beautifully adapted for this purpose.

The marks nearest home, such as the colour and situation of the hive and probably of trees in the wood, are additional to the landing signal which the bees give by standing on the landing-board facing the entrance and lifting their abdomens in the air with the scent-gland working vigorously. In addition, the supersonic signals are also sent from the entrance.

The number of bees engaged in signalling is especially great in early spring, during the first flights after wintering, and after the swarm has settled in a new place to which the bees have not yet grown accustomed.

But there are dozens of hives in an apiary, each sending its scent and supersonic signals. How do bees distinguish the signals of their own colonies? The idea is current among bee-keepers that each colony signals to its own bees and that each has its individual "family" voice.

Let us stop to consider this assertion.

If a frame covered with bees (there may be a thousand or even more bees on a comb) is put into a strange hive, more often than not all the bees will be ruthlessly killed or driven away. A few moments after the strangers have been introduced into the hive, an angry buzzing may be heard within it and its inhabitants will be seen fiercely ejecting the intruders.

Very little is known as to how bees belonging to one colony know each other and by what signs they distinguish their sisters from strangers. But that they do so is quite evident.

A group of students of the Timiryazev Agricultural Academy were instructed to carry out a simple experiment. They put 300 bees in a box divided into three compartments by a wire net so that the first two compartments contained each a hundred bees belonging to one colony and the third, a hundred bees from a strange colony. Only bees in the middle compartment had access to syrup and water. They busily sucked syrup and passed it both to their own sisters and to the strangers. It was quite plain that the former were served first and came in for a larger share. Twenty-four hours after the experiment was started the bees which had died of hunger were counted and it was seen that 60 bees had perished in the strangers' compartment while in the sisters' compartment there were 22 dead bees—nearly a third. The bees which the circumstances had turned into nurses ceased to distinguish between their "own" and "strange" bees not earlier than on the second or third day after the beginning of the experiment.

The physiological proximity of bees belonging to different colonies manifests itself in the ease with which they feed one another.

The process of "assimilation" proceeded differently with different races of bees, revealing new aspects of the phenomena of selectiveness, affinity, and compatibility. But when bees from one colony are introduced into the hive of another, incompatibility may manifest itself in the annihilation of all the strangers.

However, this intolerance, which some biologists vainly tried to represent as a manifestation of intravarietal struggle, is not an immutable law of bee life. When necessity arises, experienced apiarists manage very well to reinforce their colonies by introducing bees from other stocks, and the bees take well to the strangers, if manipulated skilfully. This "grafting" is especially successful during an abundant honey-flow.

Neither is there anything unusual in grey Caucasian bees being peacefully accepted by black forest bees, or Ukrainian steppe bees by big yellow Caucasians.

It often happens that when individual strange bees come to a hive loaded with nectar of pollen they are willingly admitted into the hive.

But the idea of a wide-spread "roaming" of bees in a closely-packed apiary is somewhat exaggerated. Although some colonies in apiaries consisting of various strains and races of bees may be made up of differently-coloured bees, this is often due not to the "immigration" of strange bees but to intravarietal cross-breeding, the progeny of which may possess various intermediate characters.

And indeed, observations at the apiary in Gorki Leninskiye show that a colony obtained through cross-breeding may consist of bees possessing not only the mixed characters of the two breeds. From neighbouring cells, out of eggs laid by one and the same queen, black, typically northern bees and bees with the broad yellow bands typical of southern strains were hatched. The offspring as a whole was heterogeneous, although each individual bee presented a pure type. Anyone observing the landing-board of this colony might think that it admitted strange bees, when actually all these bees were full sisters, members of one colony.

It is quite natural that bees should be attached to their hive. Field-bees provide the whole of the colony with food which enables its members to produce wax—the building material for comb construction. Field-bees are the only members of the colony that produce means of life, the rest being engaged in working the raw materials brought by the foragers and in reproducing the colony's population. The stable instinct attaching field-bees to their hive may be considered as another manifestation of the usefulness of biological adaptations fostered by natural selection.

But it is not worker-bees alone that can find their home and distinguish their own hive from strange hives.

The queen, too, has this instinct highly developed. Apiarists engaged in queen-rearing can tell anyone that if a queen kept in a cage happens to escape from it during manipulation, the apiarist has only to wait patiently without changing his position, and the queen may soon return and enter the cage in his hand. Drones, too, contrary to what is usually said about them, return from their trips to their own hive. If a hive is carried at night some five kilometres away from its place and put in a locality beyond the former flight range of the bees, the worker-bees will be unable to return to the old place. But unlike the worker-bees which will come home to the new place at nightfall, the drones will be found on the stakes of the stand from which the hive has been removed.

This may lead to the belief that drones are attached to the spot where the hive is even more strongly than worker-bees. But the important thing is not the force of attachment but the radius of flights: drones fly from their hive much farther than workers and, consequently, can return home from greater distances.

For millions of years their nest remained stationary for bees living in rocks or hollow trees, and this explains why the majority of foragers will alight on the spot where the landing-board used to be if, during flight hours, the hive is turned facing another direction than the accustomed one. And if the hive is moved somewhat aside, almost all the bees will fly to its old place, no matter how well the hive may be visible. They will have to make an extra effort to be able finally to reach home.

All features of bee behaviour show that bees are firmly attached to the location of their hive and that the various colour landmarks, supersonic and scent signals are but auxiliary landmarks on their way home.

Temperature and Humidity in the Hive. How Honey Is
Produced. More about Physiological Barriers. What
Technologists, Histologists and Biologists Say about
Honey. The Sting: the Colony's Weapon of Defence.
The Modification of the Sting through Natural Selec-
tion.

Nectar has been described in literature as "the soul
of the flower," "a mother's smile," "life's touching striving
after happiness and beauty," and so on. We must, however, admit that this "mother's smile" is insipid enough,
for the nectar of flowers contains from 40 to 80% water,
so that the bees have to evaporate half or three quarters
of what they bring into the hive. Ripe honey must contain not more than 20% water.

Gogol's bee-keeper Rudy Panko (in *Evenings on a Farm-
stead near Dikanka*) was perfectly justified in speaking
of ripe honey that nothing more delicious could be found
anywhere: "I will take my oath on that. Just fancy, when
you bring in the comb the scent in the room is something
you can't imagine; it is clear as a tear or a costly crystal
such as you see in ear-rings."

However much honey a colony may have stored it will
not stop collecting it as long as the flowers have nectar
and there are empty cells to store it in. This insatiate
and insatiable urge to store food has given rise to an apt
saying: "Bees are misers—they store up honey and then
die."

The warmth of the sun accelerates the process of evaporation of nectar in the cells, but the greater part of the
work is done by the bees themselves. And it involves a
considerable expenditure of energy: 536 minor calories
are required to convert a gramme of water into vapour.
But an average bee-colony consumes 90 kilogrammes of
honey a year, for the production of which 400 kilogrammes of nectar must be brought into the hive! In addition,
bees produce extra honey, too, and it is quite usual in
the U.S.S.R. to harvest as much as forty to sixty kilogrammes of honey from a colony. How much water must
be evaporated from nectar in such highly-productive

colonies if average stocks storing not more than 50 kilogrammes of honey against the winter evaporate hundredweights of water.

This living nectar-refinery processes the raw material it itself provides, and the final product is at the same time the fuel used in the process. We know the efficiency of this fuel: bees must consume about two kilogrammes of honey to produce ten kilogrammes of honey out of nectar.

Nectar must be concentrated for several reasons. In the first place, thin honey could not be kept for any length of time and cells containing it would become little fermentation vats. Secondly, less space is needed for storing concentrated solutions.

The hive, however, is not merely a nectar-refinery and honey is not simple nectar sufficiently de-hydrated. Evaporation of moisture from nectar is only one aspect of the honey-ripening process: there is another, a much finer, side. Technologists consider it a technical miracle that the eighty-per-cent sugar solution produced by bees can keep for years without granulating.

The ripening of this over-saturated sugar solution starts in the bee's honey-stomach on her way home from the field. There, the cane sugar of the nectar, under the influence of enzymes—ferments excreted by certain glands —is partially changed into a mixture of equal quantities of dextrose and levulose.

Once in the hive, the nectar-gatherer either herself deposits her load in a cell, crawling into it upside down and attaching a hanging drop to the roof of the cell so that it runs slowly down the side walls (a little drop will dry the quicker), or transfers it to a home-bee. Bees meandering over combs and seeming to do nothing are just home-bees which have received loads of nectar and are continuing the process of ripening begun by the fielders.

After this second procedure the somewhat changed nectar is put temporarily into cells to be removed to higher-lying cells some time later.

The above details are important not only from the technological standpoint: they supply one more link in the chain of metabolic processes occurring in the bee-colony, where every drop of honey and every scale of wax is the product of common effort.

We described above how a laying queen is fed with royal jelly—the substance out of which eggs, the embryos of future bee generations, are made in the queen's body. Then we traced the ways of producing honey and beebread —substances which are converted into royal jelly in the bodies of workers. We also know how bees collect nectar and pollen out of which honey and beebread are produced. This enabled us to follow, one after another, all the phases in the process of the development of sex-cells.

This process clearly and convincingly illustrates the correctness of a major premise of Michurin agrobiology: like all other cells by means of which organisms can be multiplied, sex-cells are the result of transformations, of metabolism, the result of the development of the organism as a whole.

These cells, which have accumulated as it were, the entire course of evolution of the organism and, consequently, all the conditions of life prompting its development, are born, built up of molecules, of particles, of substances constituting various parts and tissues of the organism that have gone through numerous but law-regulated changes. In the words of Academician Lysenko: "Hence, the initial cells also express to a greater or smaller extent the tendency of the future properties of the organism."

This is exactly what can be observed in the bee-colony as a whole. All its members have a direct or indirect share in the production of sex, reproductive, cells which are built up out of substances transformed in accordance with natural laws and processed in a strict order by bees of various age groups. That is why the sex-cells possess certain hereditary tendencies.

The details of honey-ripening just described reveal one more aspect of the adaptations making for conservatism in bee heredity.

We have mentioned nurse-bees forming part of the retinue which, like an inner filter, surrounds the queen and protects the reproduction process of the colony from an excess or lack of food or some of its components. Now let us remind our readers that the nurse-bees producing royal jelly for the younger larvae are protected from direct influences of the environment by the very nature of the colony much more rigorously than it would seem at first glance. And indeed, the nurse-bees produce the larval food out of food that has been processed to a certain extent, pre-fabricated, so to say, by the nectar-and pollen-gatherers and the home-bees that deposited the loads into combs. At this stage the qualitative differences of the raw initial products are already eliminated.

The fact that nurse-bees are, as a rule, young bees that do not leave the hive and are still in the "waxen womb" is also of some importance. It is but natural that the food produced for the brood by the nurse-bees living under conditions of constant temperature and humidity and eating uniform food, is more uniform, qualitatively constant so that through its consumption constant characters and properties are developed in the brood.

The peculiarities of the alimentary canal of the honey-bee also act in this direction. Every grain of food that passes from the honey-crop into that part of the canal where food is digested for the benefit of the individual bee is instantly enveloped by the cells of the peritrophic membrane, and it is in a solid covering of these assimilating cells that food makes its passage through the body of the bee.

It was always held that the importance of the cells of the peritrophic membrane consisted in protecting the inside of the intestines from the coarse integuments of pollen-grains, indigestible for bees. But there is every reason to suppose that these diminutive "wandering stomachs" are instruments of biological selectivity for protecting the heredity of bees against unsuitable food. Here is an example to bear this out: when the field-bees bring into the hive large quantities of poisonous nectar or pollen (such cases do happen) it is young bees that suffer most and subsequently perish from this unwholesome food. Since each young bee feeds several young larvae, the death of one bee preserves all the larvae she feeds.

We see how numerous are the protective adaptations

with which nature surrounds the growing brood, and especially the queen. A man that would change the nature of bees has before him the hard task of overcoming all these living and enduring obstacles, filters, and barriers.

In bringing closer to their dwellings cut hollow trees containing bees, men at the same time transfer the environment which bees create for themselves in their nests and which is the prime condition for preserving their nature, heredity. It is this environment that enables bees upon domestication to feed independently and in all other respects to lead an independent life in accordance with all natural forms and means for preserving unity with the conditions of existence.

But is the stability of heredity protected by nature less vigorously in other organisms? Are other organisms not provided with all such, or similar, filters, barriers and the like protective adaptations?

Such adaptations are certainly possessed by all living beings whose nature vigorously resists the inclusion in the development of the organism of any strange, alien conditions. But at the same time almost any living being can consume its natural food in quantities often many times exceeding that which is necessary for the normal growth and development of the organism. Excessive feeding undermines heredity and no matter how well the reproductive organs are protected, the consequences of excessive feeding will influence them, directly or indirectly. Too much fat may make an organism sterile. Good crops are never collected from soils into which too much fertilizer has been introduced. Trees that receive too much food produce sterile, fatty shoots. An overfed bull or boar cannot be a good sire. A fat hen does not lay well.

Moderate fattiness, however, results in changing the offspring, and plant- and stock-breeders have been long utilizing this for selection purposes.

Here is how Darwin ends his *The Variability of Animals And Plants under Domestication*:

"A habitual excess of highly-nutritious food, or an excess relatively to the wear and tear of the organization from exercise, is a powerful exciting cause of variability."

In bees, however, the anatomy and instincts of individual insects, as well as the organization and way of life of the colony as a whole, are such as practically to preclude an excess of food relatively to the wear and tear of the organism of individual insects from exercise.

Whatever the function each individual bee may be performing at a given moment, she usually consumes no more food than is necessary for the production of the corresponding quantity of physiological energy.

Bees are capable of incessantly carrying into the hive an abundance of nectar of the very best quality, but an individual bee will, as before, take for its individual consumption just enough food to be able to live and work.

This peculiarity which bees share with other social insects and which has enabled the honey-bee to serve man, constitutes another adaptation for the preservation of the stability of hereditary characters both in individual bees and in the colony.

It should be mentioned here that a grown bee eats very little. When the food consumed by an adult bee during her lifetime was divided into food for the continuation of her individual life processes and food for performing various duties in the colony, it was established that during the six weeks of her adult life, a bee consumes little more food for the former purpose than during a week of her larval life. If, as we have already had occasion to mention, the larval phase is considered a phase of accumulation of the individual's food store, and the pupal stage, that of consuming this food, an adult bee may well be considered to be in a phase of accumulation of stores for the colony.

Let us now go back to the hive where fanning bees are busy driving from the hive the air laden with vapours from the "green," or unripe, honey filling the combs to the brim. During one night, the volume of the nectar deposited in the combs decreases almost by a quarter: tireless fanners force with their active wings myriads of vapour molecules out of the hive. At last the cell on the upper wall of which a single transparent drop of nectar sparkled but a few days ago, is filled with a thick, gleaming liquid to be preserved.

This liquid consists of 80% sugar, a negligible percentage of salts, vitamins, ferments, an admixture of pollen, a little protein, traces of several acids, indefinable substances contributing to colour, to aroma and flavour, and some other substances. And all this together makes honey. As soon as the honey is ripe the bees seal the cells with wax cappings, each strain in its own way.

Different sources yield different honeys, such as golden-yellow from the locust-wood, white granulated honey from the yellow acacia, reddish heather honey, dark buckwheat honey, light-amber honey from the linden, sweet clover and sunflower, and white from the willow-herb. Natural honey straight from the combs is more or less liquid and viscous. Heather honey alone is thixotropic (like jelly) and special devices are used to extract it from the comb.

Among various honeys there is one which Maxim Gorky mentions in one of his stories. It is called "heady honey" and may be found in hollows of old beeches and lindens. Pompey the Great's soldiers were nearly lost when, having tasted this honey, a whole legion of iron Romans became unable to move. This honey is produced from bay and azalea nectars.

The food and caloric values of honey have been studied exhaustively, and it has been established that its caloricity is higher than that of cream, caviar, and rice while it would be hard to find any other food as readily absorbed as honey.

From ancient times honey has been believed to possess special curative properties. In the works of ancient physicians it was called the "elixir of youth," and modern doctors consider it a food for longevity and often quote ninety-year-old Pythagoras for saying that he would not have lived so long had he not eaten honey. Special works on age and longevity say that a large percentage of those who have lived over 100 years are mountain shepherds and bee-keepers.

Every year brings more and more books on the technological and food properties of honey, treating the problems with increasing thoroughness. Only its biological properties have been left in the shade until recently.

Experiments in accelerating the healing of wounds with honey, the work of Soviet specialists who have discovered in honey what are called growth substances and established that plant cuttings treated with honey take root much better, show what unexpected discoveries are in store for the research worker.

But honey is interesting not from this aspect alone.

There is unlimited evidence that different colonies of bees living side by side and collecting nectar supposedly from the same plants produce somewhat different honeys. The teaching of Michurin sheds new light on this fact: the analysis of the data obtained through the study of vegetative hybridization proves beyond doubt that plastic substances, too, carry varietal properties, i.e., heredity.

Since honey is an organic food consumed by the bee-colony throughout its life, it must of necessity be different in different colonies. Take two colonies of equal strength living side by side in the crown of the same lime-tree: the honeys produced by these two colonies will have comparatively different tastes, colours, thicknesses and flavours. In this we see the manifestation of the different heredities of the colonies which differently assimilate the nectar they collect. The nectars collected by different colonies, too, differ to a certain degree, for bees collect them from different plant species.

If pollen traps are put in bee-entrances every day and the botanical composition of the pollen loads is compared, it will be seen that no two colonies bring in identical loads of pollen.

Everybody is familiar with the curative properties of honey, but not everybody knows that pollen in the form of beebread is also beneficial to health.

Thus, young plants treated with colchicum preparation cease to develop normally. A thousand rye seedlings treated with this preparation developed tumours and became so many plant cripples. But when the experimenters treated a thousand seedlings first with colchicum and then with a liquid beebread extract, 600 to 700 plants out of the thousand continued to develop normally.

In a laboratory, three groups of mice susceptible to cancer were reared under similar conditions, the only difference being in their diets: group I received beebread, group II, fresh pollen and group III made up of control-specimens received a usual diet. After all the mice were infected with cancer through inoculation, many of those in groups II and III contracted the disease and subsequently died, but there were very few sick mice in group I which was fed on beebread.

The hopeful results of these experiments have started a series of researches conducted not only by biologists and veterinaries, but also by doctors.

Doctors utilize bee-venom, which has also been proved to possess curative properties. To-day, in addition to honey and beeswax, bee-venom is extracted on a large scale.

When a bee stings she introduces into the wound only about 0.0003 gramme of venom, but this minute quantity is enough to cause the death of many insects stung by the bee.

Bee-sting poison is the secretion of two

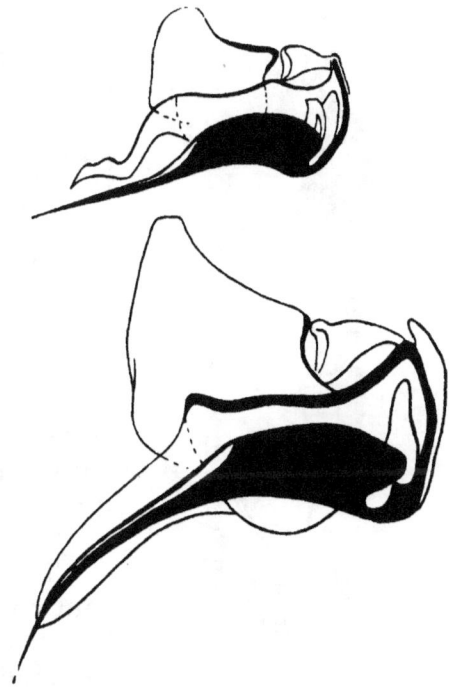

These drawings by Professor G. A. Kozhevnikov show the differences in the structure of the stings of worker (above) and queen (below)

glands, the product of one of these being acid, that of the other, alkaline. Taken separately, each of the secretions is less potent than in a mixture. A fly stung by a bee dies instantly, but if one of the secretions is injected into a fly with a very fine needle, the fly will survive. No sooner, however, does the fly receive an injection of the second secretion than it dies.

Specialists have established that some of the components of bee-venom are similar to those of the adder and the cobra venom. It is therefore no wonder that even large animals and birds often die from bee-poison. Dogs, and especially horses, suffer severely from bee-stings. Some people, too, are very sensitive to stings.

At the same time bee-venom is successfully used in treating rheumatism, neuralgia, and other diseases of man. Today the treatment is effected not only through the direct application of stings, as was the case in former days: preparations made from bee-venom are injected subcutaneously. To manufacture these preparations, special "pharmaceutic apiaries" are maintained. Here, bees are from time to time released in special glass chambers where, under the influence of ether vapours, they extrude their stings. The walls of the chamber become covered with droplets of the poison from thousands of stings. The venom is collected and bottled in ampules.

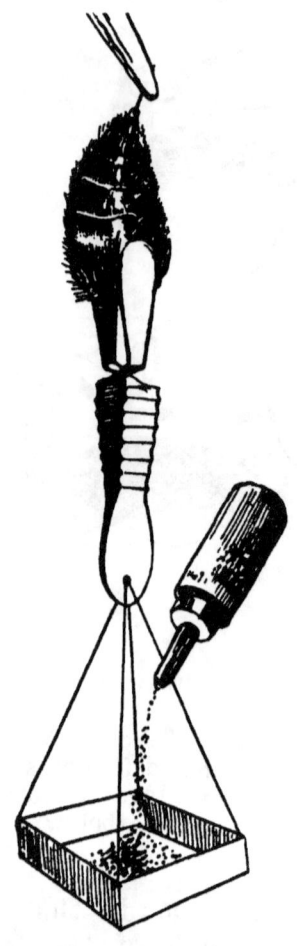

The bee-sting is held with pincers; a paper box is suspended from the separated abdomen by a very light spring clasp. By pouring sand into the box the force with which the sting is attached to the insect's body can be measured

The laboratories of the pharmaceutic apiaries have established that the maximum bee-sting poison is secreted by bees more than ten days old and that bees that consume food with a greater protein content secrete more poison. Although the magnitudes here are expressed in terms of fractions of a milligramme, still they have their relative values.

Conditions under which bees secrete more poison are already known and "exercising" poison-glands is now carried out.

It is a wide-spread notion that a bee dies immediately she loses her sting, but actually a bee without a sting can live several days. At

240

any rate, bees with their stings torn off can often make their way home and pass on their loads of nectar or pollen to house-bees.

There is no doubt that the sting is the bee's weapon of defence and not of attack.

When a bee inserts her sting into the body of an enemy, the odour of the poison infuriates other bees; the bigger and more dangerous the enemy the more defenders of the hive sting him attracting still more bees to the spot by this soundless alarm signal.

No other creature in the world has such a strange weapon of defence which in stinging an enemy often kills the stinger so that, with bees, self-defence often becomes suicide.

The sharp barbs of the sting are recurved and while allowing the bee to insert the stylet into the body of an enemy they prevent her from drawing it out, and so the bee must tear herself from her sting which, together with a bundle of muscles, remains in the wound.

When a lizard escapes from an enemy leaving behind its tail, which by its quivering holds the attention of the fooled enemy, the lizard itself in no way suffers from the loss of the tail. A daddy-longlegs caught by a leg leaves it behind and thus saves its life.

But a bee tearing herself from her barbed sting in the enemy's body receives a mortal wound.

In the chapter dealing with the difficulties in the theory of natural selection, Darwin treats this problem at some length.

The venom enters the wound down the poison canal in the ten-barbed stylet. The stylet is held firmly in the lancets (also bearing barbs) but in such a way that it can move. The next figure shows the stylet and the lancets in different positions

16—9   *241*

"If our reason leads us to admire with enthusiasm a multitude of inimitable contrivances in nature, this same reason tells us, though we may easily err on both sides, that some other contrivances are less perfect. Can we consider the sting of the bee as perfect? . . .

"If we look at the sting of the bee, as having existed in a remote progenitor, as a boring and serrated instrument, like that in so many members of the same great order, and that it has since been modified, but not perfected for its present purpose, with the poison originally adapted for some other object, such as to produce galls, since intensified, we can perhaps understand how it is that the use of the sting should so often cause the insect's own death; for if on the whole the power of stinging be useful to the social community, it will fulfil all the requirements of natural selection, though it may cause the death of some few members."

It may be noted here that Darwin's idea of the original uses of the poison was not mere speculation. Darwin mentions in one of his letters that by introducing wasp-poison into plant tissues he obtained in several cases thickening and solidification of the tissue.

The works of Academician E. N. Pavlovsky present deep, all round study of the evolution of the bee-sting. All the

---

The stylet and the lancets by alternate thrusts sink deeper and deeper into the wound holding each gain by the recurved barbs along their sides. The action of the sting continues even when the sting is separated from the body of the bee and the venom continues entering the wound

aspects of this evolution, such as the comparatively greater development of the acid gland in the Osmia and of the alkaline gland in the Andrena, have been studied by Academician Pavlovsky. He supplies exact data on the properties of the poison of such wasps as the Scolia which, through stinging, paralyses and preserves other insects for its grub to feed upon. The author offers an exhaustive study of the differences in the stings of different solitary bee-species whose stings are their weapons of individual defence, and, finally, shows in what way, in primitive social bee-species, the sting of an individual became the weapon of collective defence, an organ for the protection of the community.

The better to understand the biology of this adaptation, we must bear in mind that in the case of insects in the same evolutionary stage as the bee, stinging is not fatal for the stinger, for the sting can be freely extracted from the chitinous test. So when a bee stings a wasp or a fly she remains alive.

The barbed sting had served the bee for many millions of years before there appeared on earth birds, animals and human beings, from whose elastic skins the barbed stylet cannot be withdrawn so that the bee must either remain on the enemy or tear herself away from both the enemy and her own sting.

It is but natural that the old sting is ill adapted to new conditions. But by the time new enemies of bees appeared on earth, the bee had ceased to be a solitary insect defending only itself: by that time an individual bee defended the whole of the colony, the nest with its thousands of inmates and communal stores of food. And this proved of the greatest importance. The sting became modified in accordance with its new function and the new conditions under which it was used.

Thus, Darwin's statement that the sting was "not perfected for its present purpose" is erroneous. Darwin, however, is not to blame for this error: at the time *The Origin of Species* was in print it was not known that the queen's curved sting differs from the straight sting of the worker not in shape alone. It was several years later that data were published establishing that while the worker's sting has ten barbs the sting of the queen has only four. A ten-barbed sting sticks more firmly in the body of an enemy and remains there longer, and is, consequently, more dangerous than a smooth one.

And this is not all: anatomists have found in the bee's sting special contrivances easing its detachment from the body. It would be interesting to study what purpose may be served by all these adaptations.

Bees attacking a bear about to loot the hive get entangled in his shaggy coat and, buzzing angrily, try to penetrate to his skin. At the first sting the bear makes an attempt to brush off the guardians of the hive. In trying to free himself of the attackers the bear crushes dozens and hundreds of bees, but their stings (which are detached with a bundle of abdominal nerves owing to which the sting muscles continue to move) will automatically penetrate deeper and deeper into his skin.

"They continue fighting even after death" were the words of a scientist who was the first to understand the mechanism motivating the detached sting.

Yes, indeed, bees may be crushed, stamped to death, they may perish, but the stings torn from their bodies will go deeper and deeper into the wounds, injecting into them the poisonous mixture of the acid and alkaline secretions, owing to which the wounded spot is first benumbed and then begins to itch.

In this we perceive an obvious perfection of the sting adapting it to its present functions, although this perfection involves the death of the stinging bees.

Actually the colony does not suffer from the death of a bee, at any rate its loss is no more dangerous than is for a nettle bush the loss of the stinging cells of its leaves.

ENEMIES OF BEES

How Bees Defend Their Hive. The Bear, the Marten, and Other Four-Footed Enemies of Bees. Birds as Enemies of Bees. Insects as Enemies of Bees. An Attack of Ants. Bee Life: Proof of the Absence of Intravarietal Struggle and Assistance.

Stores of honey and wax, the semi-darkness and humid warmth of the hive, have made it attractive for a variety of enemies from parasitic moulds to birds of prey, and from the braula coeca to the bear.

In hot sunny spells when the bees are vigorously fanning the hive and when the warm smell of honey and wax floats over the apiary and is perceptible even to man, whose sense of smell is much weaker than that of animals, one may observe with interest the goings-on in front of the bee-entrance. A colony of medium strength would serve this purpose best.

On the alighting-board the hive's vigilant guards are actively repelling untold hosts of enemies. Here are all kinds of flies, big and small, long-legged spiders, slender bright-coloured moths, agile ants trying to get into the hive unseen in the crowd of home-coming bees. Now and then in front of the hive occur swift engagements, on land and in the air.

Often a live ball rolls out of the hive and falls in the grass near the landing-board: these are the inner guards driving out a foe from the hive. Long after this, an angry buzzing may be heard in the grass strewn with dozens of robbers badly wounded or stung to death.

Their place at the entrance is taken by new freebooters doing their utmost to get at the sweet riches of the hive.

Chapters dealing with the enemies of bees in books and manuals on bee-keeping usually do not mention these "border conflicts." When bee-keepers speak about the enemies of bees, they mean more dangerous foes whose harmful deeds are done by stealth in the hive: microbes, bacilli, microscopic ticks, and also specialized enemies preying on brood, damaging combs, devouring beebread, such as the wax-moth, the Phora incrassata, the Trichodes apiarius, and the like.

Besides these, bees and their homes have countless other enemies whose ways and means of attacking them are immensely varied.

The bear (whose very name *medved* means in Russian "lover of honey") is a well-known enemy of bees. The bear is attracted to beehives not so much by the smell of honey as by the buzzing within hollow trees. Men engaged in maintaining telegraph wires strung in forests have often seen Bruin climbing a telegraph-pole believing the droning sound of the wires to be the buzzing of bees.

This is how old engravings represented Bruin getting at honey

When a bear has discovered a bee-tree, he gnaws at the tree, breaks the hollow with his paws and grabs the combs paying little attention to bees attacking him on all sides.

Incidentally, when bees protect their hive from man or beast they usually start by stinging the eyes, the nose and the lips; this manifests their "tactical instinct" developed through age-long war against the bear whose whole body with the exception of these parts is covered all over with a thick fur. The bear, however, is not the only quadruped lover of honey.

A trapper in a forest in Vozhega District, Vologda Region, followed a marten's track to an old aspen with a hollow situated at the root of the tree. To drive the marten out of its home, the man began tapping on the tree. Suddenly bees came falling from above on to the snow where they froze instantly. The next day the man came to the tree to drive out the bees and take away the honey. In emptying the bee-hollow he saw that the marten had made a tunnel from its own hole to the "upper storey" and used to visit the bees for a taste of honey.

Sables and wolverenes are known to like honey, too.

Skunks are fond of eating bees. They are insensitive to stinging and come at night to the apiary, scratch at the front of a hive and when the bees appear, kill them with their paws and eat them.

The hedgehog is so impervious to bee-venom that serum on hedgehog blood may be used as an antidote against it.

In winter-time when bees are all in a cluster and no guards are posted, mice make inroads not only on wild bees, as does the marten, but also on hives in apiaries. Mice are attracted not by honey, a carbohydrate, but by beebread, a protein food. They devour dead bees fallen from the cluster and, availing themselves of the fact that bees cannot leave the cluster for fear of dying of cold, attack combs of beebread which they gnaw up to the spot where the cluster begins. Sometimes mice eat live bees from the cluster, trying, however, to avoid touching the abdomen.

Even in summer, with the bees active, mice try to get into the hive, attracted by the pleasant smell. But no sooner has a mouse got into the hive than it is stung by a mass of bees. The dead mouse is covered all over with propolis and nothing is left of it but a lump remotely resembling the animal.

Birds, too, do much harm to bees. Bees are widely scattered about the forage area, but the nearer to the hive the closer the bee tracks, and around the hive the air is filled with hundreds of bees. It is here that the feathered predators await their prey. The mosquito hawk (Falco apivorus) catches a bee across the body, bites off the abdomen with the sting and swallows the rest.

The blue tit acts like the skunk. In summer when there is plenty of other food, it does not molest bees much, but in winter it flies to a hive and taps on its side with its beak. Then follows a curious zoological pantomime based on the conflict between the hunting instinct of the tit and the protective reflex of the bees. No sooner does a guard-bee appear at the entrance to see what is the cause of the tapping than the tit throws her on to the snow and then carries her to a near-by tree to make a meal of the bee, after which the bird again flies to the hive and resumes its treacherous tapping.

These birds are but a few of the feathered enemies of bees. They are: pike-finch, blue tit, bee-eater, great grey shrike

The stork hunts differently. It seldom approaches the apiary but eats enormous quantities of bees in the meadow. Hundreds of bees eaten off flowers have been found in storks' stomachs.

As to the swallow, it is not afraid of operating in the vicinity of hives, owing probably to its higher speed in flight. Quick as lightning, it swoops up, catching and carrying away bees, sometimes right off the alighting-board. But now and then the swallow has a hard time of it, and attacked by a mass of bees, it has to floe.

First among the major bird-enemies of bees is the Merops apiaster which does not molest bees going afield but incessantly attacks pollen-carriers, so that when an apiary is besieged by several such birds, the bees may stop leaving their hives altogether.

The shrike (Lanius coleurio) causes a great deal of harm to bees. Besides devouring bees to satisfy its hunger, it stores them for future use by sticking dozens of bees on thorns, pointed twigs, and the like.

Bees, however, have the greatest number of enemies among insects.

The Philanthus triangulum, a wasp with a yellow abdomen, justly called the "bee-wolf," paralyses foragers with its sting, then presses them against its body and sucks honey-crop dry of nectar or honey. Then the wasp carries its victim to a burrow almost half a metre deep and lays an egg in the thorax of the robbed bee. On hatching from the egg the Philanthus larva will eat up the dried body.

A new enemy recently discovered at apiaries in the southern districts is the Sentoainia conica, a grey fly differing from the house-fly by the white band on its head and its somewhat more slender abdomen covered with fuzz. On sunny days Sentoanias congregate on the roofs of hives where they remain motionless waiting for bees. They attack bees a-wing in order to lay eggs in their bodies. There are

The Philanthus squeezes the honey-crop of the bee it has caught and sucks dry its sweet content; then it buries the bee's body in a burrow for its larva to feed upon

several descriptions in bee journals of the air-engagements between bees and the more quick-winged flies. Bee-keepers have a simple means of defeating these new air pirates: moving the apiary into the field, meadow or steppe where the Sentoainia cannot live.

The mantis catches bees on flowers. The wingless braula coeca damages the combs. Spiders catch bees a-wing into their nets. . . . Many, many are the enemies of bees in the insect kingdom

The wingless and (it seems) eyeless Braula coeca is much harder to get rid of. This fly, resembling a louse, is commonly known as the bee-louse and lives on bees in the hive infesting mostly queens. Unlike the louse on warm-blooded animals, the bee-louse feeds not on the blood but on the nectar or honey which it extracts from the mouth-parts of its host. The six-footed insect tickles the maxilla of the bee until she sticks out her proboscis from which the parasite sucks the honey.

Even spiders hunt bees. They spin their webs across the tracks of foragers, in which sometimes even queens get entangled.

The triangulin of the Meloe variegatus lies in wait for bees in flowers, and sucks itself fast to a forager thus penetrating with her into the hive. Other voracious insects hide among leaves to ambush bees. The ugly Asilus, the elegant dragon-fly, the agile wasp and the ponderous hornet—all catch honey-laden foragers on the wing and devour them.

The hornet often attacks bees on flowers; it bites open the bee's honey-crop and sucks out the nectar. Hornets are known to build their nests near apiaries and to attack field-bees in a body. During swarming, these bright-yellow insects fly among the bees, catch one bee after another and eat them up with an amazing swiftness.

Wasps are attracted by the smell of honey and often try to get into the hive, but guards at the entrance repulse them

Many wasp species attack bees within the hive as does the Acherontia atropos or death's head moth which gains admittance into the hive by imitating the piping of the queen. At this sound the bees become immobile and the moth penetrates deep to the food stores where it may sip as much as a spoonful of honey at a time, almost its own weight.

It is worthy of notice that when there are great numbers of Acherontia the bees barricade the entrances with wax and propolis against the night.

Such defensive actions of a colony may assume even more striking forms. In his *Ussuri Territory Diary*, V. K. Arsenyev gives the following description of a truly great battle between bees and their enemies.

"Almost all the colony was outside the hive. The entrance was low among the roots which were closely interwoven on the sunny side and formed a sloping surface. The bees crowded in a dense mass at the entrance and were faced by an equally dense mass of black ants. It was a curious sight to watch the two hostile forces facing each other, neither attempting an attack. Scout ants were scurrying to and fro on the flanks. Bees fell on them from the air and the ants sat on their abdomens and, opening their jaws wide, offered a fierce resistance. Now and again the ants would attempt a flanking manoeuvre to attack the bees from the rear, but this was immediately discovered by winged scouts, and a certain number of bees would move to block the enemy's way."

The description might have ended here but the author and some Cossacks of his detachment poured boiling water over the ants and unwittingly frightened the bees away. "The bees were all instantly in the air and you should have seen how the Cossacks ran for their very lives!"

The death's head moth can suck up as much as a teaspoonful of honey if it succeeds in getting into the hive

So we see that a bee-colony is capable of building barricades against the Acherontia atropos and giving battle to ants. At the same time there are numerous instances showing the "mutual indifference" of bees, as some old bee-men termed it. This trait of bee behaviour may be observed in acts not necessarily connected with the defence of the hive.

Thousands of foragers will fly along the aerial route between the apiary and the field past a bee buzzing plaintively in a spider's web without a single one changing her course to help the unfortunate insect out. Hundreds of bees may be observed filling their honey-crops with water from a pool in which one of their fellows is drowning without one of them going to her help.

If a small spot of the landing-board in front of the very entrance is covered with gum and a few bees get stuck in it, they will try with all their might to free themselves, but the others will keep away from the dangerous spot and, as in former instances, not one will try to save the trapped bees.

True, usually more than one bee takes part in a skirmish with a fly or a wasp, and in the case of the Cetonia as many as five or six bees. But no matter how fierce the battle, it is invariably characterized by two features: other members of the colony avoid the spot of the battle and each of the fighting bees acts on her own and is often in the way of the others. While one bee is trying to drag the wasp beyond the entrance, another does her best to drag it into the hive, a third wants to sting the enemy and a fourth shakes it so that the sting misses the victim.

In the case of an enemy offering strong resistance this lack of co-ordination on the part of bees may seem the result of chance causes, therefore we shall offer a more convincing example to illustrate our point.

If an immobile bumble-bee, wild-bee or hornet larva or pupa is put into a beehive, bees will be seen to attack it with uncontrollable fury. Dozens of bees at once will bite through the soft integuments and suck up the liquid substance of the larva. In a few moments nothing will be left of it but an empty skin which several bees will be seen dragging in different directions: one towards the entrance, another towards the back wall, a third up the combs. The rest will be passing to and fro paying no attention to what is going on.

A dragon-fly easily gets the better of a bee, then, with its strong mandibles it tears its victim to pieces and devours it, leaving nothing but the wings

What is the explanation of all these facts?

To find an answer, let us analyse a battle at the entrance. It is but natural that the majority of bees witnessing a conflict pay no attention to it. Young bees leaving the hive for their play-flights or on their way to receive nectar from foragers are not of the age for guard-duty. Instinct urges foragers carrying their loads to dispose of them first of all. Thus we see that but few out of all the bees at the entrance can take part in defending the hive, the rest following under normal conditions their normal course and performing functions in accordance with their physiological ages. As to the behaviour of guard-bees and disengaged foragers, it may serve as a good object lesson in relations within a species in nature.

The teaching of Michurin has given convincing proof that in nature there is neither struggle nor assistance between individuals within a species. Bee biology proves this conclusion of Michurinist science—so very important theoretically and practically—by an exceedingly original and convincing example.

Some biologists considering intravarietal competition the basis and mainspring of evolution, did their best to prove that the bee-colony, too, is the result of the mutual struggle of bees, the competition of drones and the ruthless war of queens. Further we shall have ample proof of the groundlessness and futility of such views. Equally groundless and futile are the teachings which hold that uninterrupted harmony and absolute perfection reigns in all living nature and in bee life particularly, and which, departing from these pious but unscientific premises, explain some phenomena of bee life as the result of an urge for mutual assistance to one's own kind prevailing throughout the animal and vegetable kingdoms.

Nature disproves these views.

If the bee fails to free herself from the web, the spider will eat her up under the very eyes of the passing foragers. If the bee does not get out of the water she will drown in front of hundreds of other water-carriers. If the bee cannot extricate her legs out of the gum she will perish on her own threshold.

The bee-colony is a living organism but it consists of individuals. Each member of the colony, while being a dependent part of the whole, is at the same time a distinct being. Although an individual bee can be considered an organic part of the whole—the colony—and of the general—the species—her relations with other similar parts of the whole and of the general offer no evidence of mutual struggle or of mutual assistance.

SWARMING SEASON

THE EMERGENCE OF THE SWARM

The Last Time the Queen Leaves the Hive. Why a Swarm
Clusters. The Experiment of a Crimean Forester. The
Swarm in the New Home. The Colony's Cycle of Devel-
opment. How a Swarm Is Formed. Can Swarming Bees
Be Substituted by Other Bees?

The bee-colony has been increasing from day to day
and its combs have been filled with more and more honey,
beebread and brood. The foragers have been busy flying
to the field and back, the builders have been constructing
combs, the nurse-bees have been feeding the older larval
every minute. Pupae have been growing in their wax cells
and young bees emerging and cleaning their cradles. Surrounded by her retinue of bees touching her with their antennae and licking her with their tongues, the queen has
been going from cell to cell laying eggs. The scavengers have
been cleaning the combs and the hive bottom; the fanners
have been driving overheated and humid air out of the hive,
standing on one spot and working their wings vigorously.
The guards have been standing a vigilant watch at the entrance round which fat drones were milling.

Did I say drones? Where have they come from? There
was not a drone in the colony when the sun woke it up in
spring. They are from the drone cells built by the young
spring bees about a month ago and filled with eggs by the
queen. Here they are, the goggle-eyed males in the midst
of an exclusively female population of which none but the
old queen has ever seen a male bee.

What prompted the bees to start rearing them? Was
it the warm spring sun or the beginning of a heavy honey-
flow?

Then wax acorn-like queen-cells appeared on the combs. Within, supplied with plenty of food, the future queens are growing apace, so that the time soon comes for them to be sealed in their royal cradles.

Soon after the first queen cell has been sealed, unrest and excitement seem to seize the colony. This usually happens in May or June, on a bright calm day with the sun hot in the sky.

The approach of the momentous event can be predicted by the state of the landing-board. Usually on hot and calm morning two currents of bees are seen moving into and out of the hive. But now the landing-board is suddenly deserted at a time most favourable for flying. Then it gradually becomes covered with bees crawling about in aimless confusion. Their number constantly increases by crowds of bees rushing out of the hive and spreading all over the board and the walls, and rising into the air. All of them are seized with a desire to fly upwards, as if these sexless workers were possessed by the mating instinct of females that has been dormant for centuries.

They fly in zigzags all over the hive, performing a wild swarming dance and filling the air with a buzzing which calls more and more bees out of the hive. Some of the foragers coming with their loads of nectar or pollen seem indifferent to all this and enter the emptying hive as usual, while others join their departing sisters even without freeing themselves of their loads.

This is the beginning of swarming, the act by which the colony is divided into two.

The old queen leaves the hive together with the swarming bees; for several days she has not been fed and has considerably lost in weight.

So the old queen leaves with the swarm. . . . Incidentally, is it correct to say that she leaves of her own accord? Does not the swarm take her along, she obediently following it? For, more often than not, the queen appears with the last bees leaving the hive.

If for some reason or other the queen cannot join the swarm, the bees outside the hive wait for her and then return home. The act of swarming is thus postponed, but the bees will make repeated attempts at leaving until the first young queen emerges and flies off with them to a new home.

Issuing with the swarm, the queen leaves the hive for the last time in her life. When she left for her mating flight the queen was young, light and agile. Now it is with difficulty that she carries her body with the embryos of countless new bees within.

The queen reaches the edge of the landing-board and rises into the air to alight a few minutes later on a branch of a near-by tree or a fence or, sometimes on a stone, and the entire swarm but a few minutes ago whirling in the air like a golden cloud settles around their queen. Sometimes a swarm issuing with a queen and beginning to form a cluster may return to the hive without any apparent reason.

Swarms thus returning usually leave the hive for good a few days later, but sometimes they destroy all the queen cells and stay in the hive from which they had tried to escape.

Why was there no swarming? What prevented the bees from settling in a new place? What influenced their subsequent behaviour?

Drones always join a swarm. What makes them follow the leaving bees? And (what is more to the point)—what business have they in the new colony which leaves behind its comfortable home for a new one? The old queen has been impregnated long ago, she does not need a drone. On entering the new home with the swarm she starts laying as soon as new cells are ready. A good swarm never builds drone cells, rearing only worker-bees as if it did not need any drones. But drones will join a swarm nevertheless.

Perhaps a bee-colony must always have a few drones of its own?

It is well known that under normal conditions of development a bee-colony rears drones in the spring and keeps them during the summer season, so that they should be available in case they are needed. It may well happen that the colony has no need for drones during this particular season if no swarm issues and no attempt is made to supersede the queen. At first glance this practice may lead one to suppose that a colony rears drones not only for itself, not only for mating with its own virgins but to provide for cross-breeding if necessary. But there are observations tending to disprove this idea. There are cases in which, for some reason or other, a colony must supersede the queen very early in the spring when normally there are no drones. In such cases it may be easily observed that contemplating a supersedure bees start with rearing drones. Drone cells are built as many days in advance of queen cells as it takes a drone longer to hatch than a queen. To be more precise, the difference in time is in accordance with the period needed by a drone to reach sexual maturity.

In the present instance the bee-colony may be likened to self-pollinating plants in which the stigmas (drones) and the pistil (queen) ripen at one and the same time. But this proves that a colony can rear drones to fertilize its queen.

Whatever the case may be, drones join a swarm, although not in large numbers: not more than a hundred, while the number of bees in a swarm varies from 10,000 to 30,000 and over.

Swarming bees are exceedingly peaceable. Perhaps the reason is that their honey-crops are so filled with honey that the physical exertion of stinging is too much for them, or, that having abandoned their old home and not founded a new one the bees are deprived of an impulse to attack.

A clustered swarm may remain motionless for hours as if waiting for the bee-keeper to hive it; this trait of bee behaviour is very valuable for man and is a character which artificial selection has automatically strengthened and continues to develop. It is natural that swarms from colonies that settled far from the hive or high up in trees or flew away soon after settling escaped from apiarists more often. Swarms which settled near the hive and low down or stayed longer on one spot before flying to a new nest were and are kept (we should be more correct in saying: remained and remain) in apiaries. And in this way varieties were formed.

Why does a swarm form a cluster?

Perhaps lest the bees should get lost and, certainly, lest honey should be wasted on useless flying. The loss

258

of weight per day in a cluster is not more than 1.5% while an individual bee may lose as much as 25%. So bees in a cluster may subsist on a given quantity of honey 9 to 10 times as long as singly.

It should be noted that bees in a swarm cluster seem to feed one another: they probably redistribute the honey taken from home.

A middle-sized, compact swarm. There are bigger swarms. A swarm weighing 2.5 or 3 kilogrammes contains 25,000 to 30,000 bees

Absconded swarms often fly a considerable distance away. Certain observations lead to the belief that, if allowed to seek their new home for themselves, swarms try to settle beyond the flight range of the parent-colony, i.e., beyond the zone they used to frequent before swarming.

How does a swarm find a new home under natural conditions?

A Crimean apiarist placed several scores of old skeps in various sequestered spots in a wood. As the swarming season came, bees appeared in many of the skeps. First a few bees came, then more.

Those were scout bees, "billeting parties," from colonies about to swarm. They occupied the skeps lest strange swarms should take possession of them. The scouts kept in constant touch with the parent-colony until the swarm came and settled down in the new place. If no swarm came, the bees lived in the skep for a considerable time and continued flying to the old hive.

The guess that scout-bees leading a swarm to its new quarters communicate to the bees its distance and direction by dancing has been proved by successful experiments carried out on a little desert island.

Swarms had nowhere to go from this island surrounded by wide expanses of water on all sides. Decoy hives were scattered all over the island and observers were posted at the entrance of each marking with paint scout-bees that came to their hives. Other observers noted the behaviour of the marked bees in the observatory hives from which swarms were expected. A third group of observers kept an eye on the marked bees in swarms that had already left the hive and clustered before proceeding farther on their way.

By the end of the first season it became clear to the research workers that while the dance was performed by scouts marked with different colours (i.e., from different places), the swarm did not leave its temporary shelter. But when the majority of scouts performed the same dance, i.e., called the swarm to the same place, the swarm followed them.

The organic unity of the colony is manifest with the greatest clarity in a swarm, a part of the colony that has become independent. A powerful force of mutual attraction gathers in a cluster, a compact mass, the tens of thousands of individuals that have left their home behind. When a swarm rises into the air to fly to its new home, even then the bees fly in a compact mass. The living cloud of bees, now thickening and now thinning, now rising high above the trees, now descending almost to the ground, the swarm, floats in the air on the way to its new abode.

Once on the chosen spot, the cloud breaks up and falls down in a live rain of bees. A thick stream pours into the entrance, sometimes but a scarcely perceptible hole, while a few are still buzzing overhead.

It goes without saying that a good bee-keeper will not let his swarms abscond—this is considered an unforgivable blunder. He usually hives his swarms in his own good time.

If a swarm has settled not too high up the bee-keeper easily shakes it into a hiving-bag or a skep and takes it to the cellar; later on — usually towards evening—the bees are driven into a hive specially prepared for the swarm. The honey brought from the old hive in the bees' honey-crops will be used for food and for secreting wax during the first days in the new home.

Crowds of bees gather on the landing-board and on the front wall before swarming

After a swarm has been hived, bees have often been noticed coming to the spot where the swarm used to be and crawling all over it. Perhaps these are the scouts we have mentioned before. If so, they are too late—the swarm has been taken away and their colony has been given a home by the bee-keeper.

A swarm may be shaken into an open hive or the bees may be released in front of one. In the latter case, too, it takes the bees not more than a few minutes to enter it. After the queen has passed through the entrance surrounded by her retinue she is followed by a stream of bees.

The bees walk at a moderate pace holding their bodies unusually stiff and, beating their wings. Their even buzzing rings as an invitation to those that follow. In half an hour the swarm will have settled in their new hive.

The bees lose no time in starting to build combs; hundreds of them make orientation flights preparing to go afield; the guards take up their posts. Soon the queen starts laying eggs in newly-built cells while other cells are being filled with fresh nectar.

No bee will now return to the old hive: the new hive has become their home while the old one, for which each of them would have laid down her life, has lost all power over the bees.

Not one individual bee, but hundreds of thousands of

bees in the swarm have relinquished the combs they themselves built, left behind the larvae and pupae they fed and the stores of honey they gathered. But the day before they knew the location of their hive so well that it could not have been moved aside a metre, or turned about, or raised a little without causing confusion among the foragers. Had that been done all the fielders coming home would have alighted exactly in the spot the hive used

Foragers back from the field often join the departing swarm with their pollen-baskets loaded

to be as if it were in its old place. Yet all of a sudden tens of thousands of bees, half of the hive's population, have forgotten all about their old home and are making themselves comfortable in a new one.

The swarm has become a colony beginning a new cycle of development.

Meanwhile life is resuming its wonted course in the old hive. Reduced by half, the parent-colony continues to grow although it has no queen as yet. Virgin queens are ripening in their sealed wax acorns and the oldest is due to emerge any minute.

Perpetual motion and renovation are observable with extreme clarity in the bee-colony where the new is ever growing and developing and the old wearing away and dying off. The colony as a whole is continuously growing and developing, and being constantly changed by conditions, presenting a spectacle of flexible, living unity with the conditions of its existence.

As the sun rides higher in the sky, beginning with January or February, the queen resumes laying, occupying cells from which honey had been eaten during the winter. The colony is waking up.

By the beginning of spring, when the first catkins—the earliest source of honey—appear on the leafless branches of willow-trees, there are very few young bees in the colony.

But in a week or a fortnight their number speedily increases; the autumn and winter hardships tell on the old bees which die one after another, and the colony is rejuvenated as if after moulting. In the meanwhile the queen is at the height of her performance, filling with eggs the combs that have become empty by this time.

More and more young bees hatch from the eggs, the population speedily increases, the colony is building up to swarming strength.

During the blossoming of the main honey plants, all foragers are busy filling with honey the combs from which the brood has emerged. The combs are filled with nectar and the queen has less space to put her eggs in, so she lays less and the population, too, becomes somewhat less, for a time.

After the main honey-flow some cells are freed of nectar and the queen uses them once more.

However, as summer draws to a close and plants secrete less nectar, the bees begin to feed the queen less frequently, thereby restraining her urge to lay. Then the honey-flow stops and the bees start driving away the drones in preparation for the winter.

Then the cold comes, forcing the reduced colony to huddle together first in an autumn cluster, then in the winter one.

This is the annual course of events in the bee-colony.

The shortest and most exciting period is the swarming time when the colony is ripe for reproduction and divides into two.

The appearance of new colonies—swarming—proceeds differently in different bee-species. Like ours, Indian bees start preparations for swarming by rearing queens and drones. At the same time, not far away, sometimes beside the hive, they start building new combs where the swarm makes its new home. Since Indian bees live in the open air, the crowns of giant trees affording them shelter become covered in the course of time with numerous bee nests, old and new, big and small.

The American Meliponas do not build a new nest beforehand but, like our bees, begin building it in a new place after swarming. Unlike the honey-bee, Melipona swarms take young queens with them and the old queen remains with the parent-colony in its old home.

Swarms of African bees consist of worker-bees only; after swarming they rear a queen out of an unfertilized egg of a laying worker. Such a queen does not live long and soon after her mating flight she is superseded by a permanent queen reared out of the temporary queen's fertilized egg.

We have had occasion to remark in connection with the experiments of M. Z. Krasnopeyev that equal quantities of food provided by different numbers of nurse-bees are qualitatively different. In that connection it was mentioned also that the biology of swarming provides convincing proof of the correctness of this statement.

And indeed, can it be a mere coincidence that for all the differences in the details of swarming in honey-bee, Indian-bee and Melipona colonies, under normal conditions new queens are reared neither by the swarm nor by the part that remains behind, but by the colony as a whole.

African bees may at first blush seem to be an exception, since here a new queen is reared by a swarm after it has settled down in a new place.

That is so, but these same Africans prove that queens reared by swarms are temporary ones and that only after a swarm has built up to colony strength does it rear a permanent queen. So here, too, true queens are reared by whole colonies.

Perhaps drones, too, like queens, are perfect for breeding purposes only when reared by a strong and populous colony?

If such be the case, there is nothing to wonder at drones joining a departing swarm. For under all conditions a swarm is at the start but *half* a colony.

Every day hundreds, and at times thousands, of new inmates emerge leaving their cells and taking their places in the routine of colony life. Withal, the colony continues to be a unity.

And then, among these thousands there appears—or is to appear—one bee, somewhat bigger than the rest, with a longer abdomen and shorter wings, devoid of wax-glands and endowed with a four-barbed curved sting. One would think her of little account, taken in the mass of her sisters. And yet, the fact that she is about to emerge causes a break in the gradual processes of growth. Her appearance brings into the life of the colony profound qualitative changes imperatively urging the colony to divide into two.

In the life of the honey-bee, one of the two colonies that have grown and developed within the parent colony takes the old queen with it and leaves for a new home.

Bee-keepers as a rule are not satisfied with this state of things, and not merely because they are afraid of losing a swarm. In the first place, the elemental, uncontrolled nature of swarming does not suit the bee-keepers' purposes. There are years when swarming occurs too often, and others when none takes place at all. Sometimes swarms are sent by weak colonies whose propagation is not in the bee-keeper's interests. And, finally, there is always the danger of losing honey as a result of swarming.

During the time a colony is preparing to swarm the bees collect half the amount of honey that is brought in by colonies not contemplating swarming. After swarming, both the parent and the new colonies spend some days in settling down to the usual way of life. Since there is no guarantee that swarming will not occur during a heavy honey-flow, the bees may lose the only opportunity of storing honey, especially in localities where the honey-flow is short. That is why the majority of apiarists regard natural swarming as nothing short of a calamity and do their best to control this elemental force. To be able to do this, one must understand what takes place in a swarming colony, what chain of events leads to swarming.

The history of attempts to answer this question might fill volumes.

Here are some of the explanations of the swarming process.

Bees swarm because they are overcrowded in an over-populous colony.

The hive becomes too small for the bees.

The combs are filled with honey and no storage room is available, so the bees have to search for a new dwelling.

The tendency to swarm is a heritage from the wild state. Domestication has weakened this tendency but has not yet eliminated it.

The cause of swarming lies in the age of the queen.

Swarming depends on the temperature in the hive.

Too warm nights make bees swarm.

The main cause of swarming is too much brood in the hive.

Old combs with cells that have become small through long use produce small bees tending to swarm. If new combs are put in every year the bees will cease swarming.

Ever new hypotheses were put forward and each collapsed, unable to stand the test of observation and experiment which, however, yielded facts shedding the first rays of light on the darkest page in the natural history of bees.

It has been established that it is not just any bees that make up a swarm.

Several hundred bees in a swarm were marked and the swarm, together with the queen, was put back into the hive. In such cases swarms usually make another attempt to leave in a few days.

And indeed, two days later the swarm left for the second time and 90 per cent of the marked bees were found in the swarm. This experiment was repeated several times with similar results, showing that the make-up of leaving swarms was on the whole constant.

Other experiments with bees marked according to their ages have shown that swarms include bees of all ages. It is quite possible that the laws governing swarming, too, are connected in some way or other with the action produced by the secretion of the queen's chitinous integument. When research has established the age of the bees that lick the

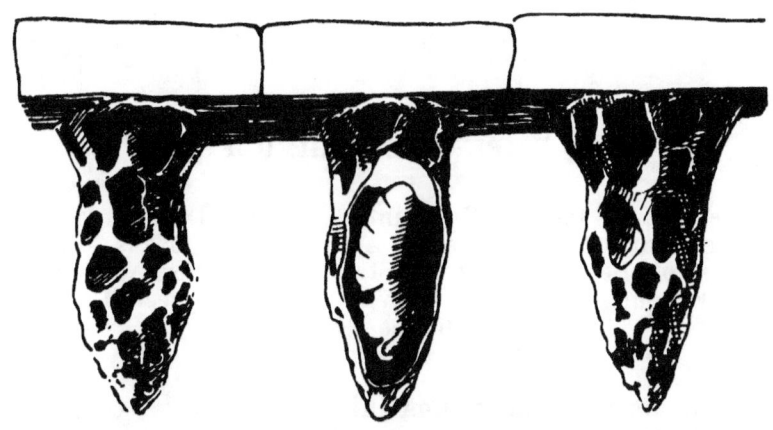

Several queens are reared at once in a queen-breeding nucleus. We see three sealed queen cells. One of these is cut open and a pupa with the end of its abdomen surrounded with the remains of unconsumed food can be seen

queen, and also the influence on the colony of the workers' touching the queen with their antennae, much that is obscure to-day will become clear.

But although all these data may be important and instructive, they supply no answer to the question uppermost in the minds of bee-students.

"It is all very well," a biologist may say. "We know how the urge of swarming takes possession of a colony. But we do not know what causes this urge. We know how a swarm is formed. The sequence of conditions taking place in a swarming colony has been followed in detail. But what starts it going? Is it the instinct of propagation? Undoubtedly! But then why in the case of two colonies living under similar conditions and in every respect alike, one sends out a swarm while the other quietly supersedes its old queen and goes on growing and assiduously collecting honey? Why out of two colonies under equal conditions one, after sending a swarm, settles down to work immediately while the other continues to send out new swarms as if torn by inner strife?

We shall devote the following section to the study of these questions, so important for the correct understanding of bee biology.

THE "PIPING" OF THE QUEENS

Intervals in Laying Eggs into Queen-Cell Cups. What "The War of the Queens" Actually Is. Queens—Prisoners of Bees. Against the Stream. Man Must Not Wait for Favours from Nature. Why the Old Queen Leaves with the Swarm.

As soon as the queen has laid an egg in a cell cup the bees finish its construction bringing it to the queen-cell stage. Upon the hatching of the larva the queen cell is always surrounded by numerous ever-changing groups of nurse-bees.

Whatever the number of cell cups, the queen never lays in all of them one after the other. The queens of the black bees cultivated in the central regions of the U.S.S.R. lay, as a rule, with intervals of from one to three days. As a result, the virgins emerge with corresponding intervals.

To appreciate to its full the value of this adaptation, one should closely observe the proceedings in a hive that has just sent off a swarm.

The number of bees in the hive has noticeably decreased. A few days must pass before the first virgin is due to emerge from the first queen cell, and the entire colony seems to be eagerly awaiting this moment.

On the eighth day after the sealing of the first queen cell a virgin emerges. She gnaws through the remains of her cocoon under the capping of her wax cradle and, throwing open the capping, emerges.

After the emergence of a young queen the development of the colony will take its natural course.

The virgin runs over the combs "piping" as if announcing her birth to the colony. This is the sound which the Acherontia atropos imitates and at hearing which bees become immobile as if paralysed.

The younger virgins, still in their cells, answer the resonant piping of their elder sister by a somewhat subdued sound, "quahk, quahk".

If the colony, after casting its prime swarm, is in no mood for continuing swarming, the bees, together with the virgin, dispose of the remaining unhatched queen-cells.

We have mentioned earlier that a queen larva is not covered completely with its cocoon: the extreme segments of the abdomen are free of it. Now we shall see what this leads to.

The cocoon is sting-proof and the unprotected segments of the abdomen are the Achilles' heel of the ripening queen. When the colony is disposing of the extra queen cells by biting through their walls, bees insert their stings into this part of the queens' anatomy. This peculiarity of the queen cocoon is a very fine biological adaptation existing evidently for the sole benefit of the colony and the species as a whole.

Biologists who observed nature through the distorting mirror of their erroneous theories saw in the destruction of the superfluous queens an instance of intravarietal struggle. But then one might be equally correct in regarding the falling off of flowers from plants or the dying off of buds as a struggle between the organs of one and the same organism, and not an adaptation to conditions and proof of the harmony governing interdependable vital processes.

Pavel Ilyich Rashevich, the main character in Chekhov's *An Estate*, considered himself an "inveterate Darwinist" just because he attached a profound significance to such words as "blood, aristocrat, pure strain." Biologists that consider the killing of superfluous queens "the war of the queens" whose purpose, they hold, is to "select the strongest and the fiercest" have an equal claim to the name of Darwinists.

It is quite clear that if such were the natural way of selecting the fittest queens for the propagation of the spe-

cies, all the queen cells would be constructed simultaneously and the virgins would emerge also at one and the same
time, or in groups, so that in battle the weakest might
die and the fittest survive to the glory of intravarietal
struggle and competition as the "basis for perfecting
species".

The *war* of the queens is pure imagination

Actually, there is no contest or battle whatever.

A naturalist who sincerely believed that two queens
meeting in the hive must engage in a deadly fight has left
us the following description of such a meeting. The queens
"took up such a position that if they used their stings each
would have received a deadly thrust at the same time....
A second later, however, the two female combatants, seized
with fear, rushed in opposite directions. When they met
some time later, they again ran away from each other
feeling that their death would be fatal for their subjects."

The description is quite true to life. But what is its
meaning?

It is true that, in the end, one queen stings the other
to death. But can we call a war, a battle an encounter in
which both sides behave as if each were afraid of the oppo-
nent's death? Do we call such behaviour war?

When a vigorous young queen and a worn-out old queen
are put together in a hive it is usually the old one that dies.
This has a certain biological significance. And there is no
doubt that further research will reveal the operation of the
law of biological selectivity in the fact that one virgin can
sting another to death.

When after sending off a prime swarm, a colony means
to cast after-swarms, the sealed queen cells remain intact
and the virgins within are allowed to mature. Since the
old queen laid eggs in them with intervals, the interval between the hatching of the virgins makes it possible for after-swarms to leave.

If the weather or some other causes prevent the cast
from leaving with the first virgin, and a second, a third,
and more virgins mature and try to emerge, the bees hold
them down and do not let them do so.

The virgin waiting for the cast to leave runs over the combs "piping" and seeming to warn her "quahking" sisters that she is still here, that the swarm has not left. And when, in spite of the warning, the virgins start getting out of their cells, the workers prevent them from emerging.

The bees not only protect these over-ripe virgins in their cells, they sometimes feed them by inserting their proboscides through the hole in the wax capping. All this lasts until the cast has left, when a new virgin may safely emerge.

Although it often happens that several virgins leave with one swarm, the above details of the normally-developing process look as if everything were done to prevent the meeting of the virgins with their highly-developed instinct of mutual intolerance.

The old, fertile, queen leaves with the prime swarm.

The virgins leaving with the after-swarms are lighter, since they carry no eggs. That is why casts and colts settle farther from the hive and higher up in trees, quickly break the cluster and take off. All after-swarms are considerably smaller than the prime and move with greater ease. Such swarms fly much farther from the hive.

Every swarm takes from the hive about half of its population, so that if the bees continue to swarm after the prime has left, the colony may become very weak.

Swarming is a manifestation of the reproductive instinct and if colonies sent no swarms the species would die away.

But if such is the case, are not bee-keepers' efforts to prevent swarming doomed to failure? Perhaps the prevention of swarming is a myopic attempt at violating the principal laws of life? If a complete control of swarming were achieved, would it not lead to the elimination of the bee-species?

By no means! All such fears are quite groundless.

Artificial swarming has long been practised by apiarists: by simply dividing colonies and shifting field-bees they may double and further increase the number of colonies at will. After division, each part grows and develops casting no swarms for some time. Experienced bee-keepers have learnt to imitate nature in this respect.

But imitating nature does not mean controlling it. It has long been known that some colonies and their progeny are more inclined to swarming while others collect more honey, so that bees may be divided into "swarming" and "honey-storing."

Many new observations have been recently added to this one.

Bee-keepers hold that, all other conditions being equal, colonies mothered by queens not reared under the swarming impulse or by queens reared late in autumn are less inclined to swarming. They believe also that queens hatched from eggs laid by older queens head colonies with a greater propensity to swarming.

Guided by these and similar observations, our expert apiarists have long been able, through strict selection and directed breeding, to weaken the tendency of bees to swarm.

P. L. Snezhnevsky, a talented bee-keeper from Orel, adopted as his motto the well-known saying of Michurin: "We must not wait for favours from Nature; we must wrest them from her" and paraphrased it in the following way: "An apiarist cannot use for breeding purposes what blind nature gives him. He must create what he needs."

Snezhnevsky could both correctly understand the tasks facing him and solve them. Directed selection enabled him to breed a big apiary of non-swarming colonies which he divided for many years as and when required. So Russian expert selectionist-apiarists have made a brilliant contribution towards controlling the nature of bees.

But it is the new understanding of the causes of swarming furnished by Michurin's theory that alone opens the way for a conscious and scientific solution of the task of subduing this elemental force at the apiaries.

Among the innumerable facts attending on swarming, the individual, the particular, conceal from view the manifestation of general laws governing all living nature.

We should, therefore, attentively study the various aspects of swarming as a particular form of reproduction of bee-colonies.

Reproduction in nature usually involves crossing. Reproduction in bee life, too, obeys the law obtaining in living nature and demanding that from time to time crossing should take place between distinct individuals.

As Academician Lysenko says, in sexual crossing by means of fertilization, "the different sex cells or their nuclei, uniting in one cell or one nucleus, produce the biological contradiction of a single living body. In this way is created a live fertilized ovule, thus it is transformed into an embryo, into an organism."

Sexual reproduction is present in the bee-colony where the queen is mated with a drone. This mating is the pledge of life, growth and development within a colony. But *colonies* are reproduced only *asexually*. The bee-colony as a biological unit is capable only of dividing, i.e., can be reproduced only vegetatively, through "self-spreading shoots."

But biology teaches that a prolonged asexual, vegetative, reproduction will of necessity result in diminishing vitality.

How is this prevented in the bee-colony?

Michurin's theory furnishes the following answer to this question. Vitality may be increased and renovation may take place not only through sexual but also through vegetative processes. Assimilation by a living organism of new conditions of environment brings about an increase of vitality.

Plant-growers know very well that it is often beneficial to sow the seeds of certain varieties in new localities with conditions differing from the old ones.

It has been established, for instance, that if barley is sown or potatoes planted repeatedly in one and the same place with their own seeds or tubers, the plants will eventually yield less. The same seeds and tubers sown in another patch, not necessarily at a great distance from the old one, may bring fair yields, while new seeds must be used on the old patch.

Scientists and practical agriculturists have long been aware of the biological advantages of slightly changing the conditions of life of plants and animals. But the facts could not be explained, and only the teaching of Michurin has discovered in them the manifestation of the general law that the biological contradictions in a living organism constitute the source of its vitality.

The study of the problems connected with in-breeding has revealed that the more a living organism is made to assimilate conditions of life selected out of the environment, the greater the intensity of the living processes taking place in the organism, the greater the vitality of the organism.

From this premise, it is clear that if generation after generation assimilates the environmental conditions for an exceedingly long period of time, or if the process of assimilation proceeds with excessive intensity, the organisms may become too well accustomed to the environment. The environment ceases to provide the organisms with conditions capable of creating the biological contradictions—the source of their vitality, or as philosophers say, of spontaneous development.

This law may explain one of the main causes of swarming.

Considered in its general aspect, swarming may be qualified as crossing with differing conditions, which in the final analysis, increases the vitality of the colony as a whole.

Depending on a variety of conditions, swarming may occur more or less frequently, one of the conditions determining its frequency and regularity being, evidently, the state of the queen.

It must not be lost sight of that the aggregate weight of the eggs a queen lays during one summer month exceeds her own weight thirty, forty times, or even more. If the eggs, each one and a half millimetres in length, laid by a queen in 24 hours were put in a row, we should get a thread two or three metres long. At the height of laying, eggs "drop" from the queen one after another for many days and nights. If no space for laying is available, a queen lets fall her eggs on her way. This may often be seen in one-frame observatory hives where the queen is cramped for space.

A comparison of the weights of the eggs a queen lays in 24 hours and the food she consumes will show that eggs are produced literally out of the food the bees give the queen. It is self-evident that such prodigious fertility is possible only as a result of highly intensive metabolism.

But the more intensively an organism assimilates the environmental conditions, the sooner its biological contradictions are smoothed out.

If it is true that, as a rule, the queen mates but once and becomes fertile for the rest of her life, the acuteness of the contradictions, the vitality possessed by the fertilized ovules must diminish in the course of time.

What may be the results of this process? A living body is viable as long as contradictions exist within it. As they become less acute, as the process of assimilation and dissimilation becomes less vigorous, the organism's vitality ebbs, the organism grows old.

These general theoretical principles help one to understand the vitality of the bee-colony.

The possible results of the ebbing vitality of the bee-colony are prevented by moving into another locality, by a slight change of conditions, which consists first of all in changing the pasture.

Forty bees in a small colony were marked with individual numbers and pollen loads on their legs analysed so as to establish the sources from which they came. Then the colony was taken eight kilometres away from the old spot and the loads of the foragers analysed again. It was seen that eight out of the forty, continued visiting the same species as before and the rest visited other plants. A forager that had worked sweet-brier began visiting hawthorn, another exchanged hawthorn for hawkweed, a third left hawkweed for dandelion, a fourth took to blackberry from dandelion.... Besides, the plants in the new locality were, of course, somewhat different.

Thus, transferring home to a new place means a change of pasture, it means a substitution of new environmental conditions for those already assimilated.

We may suppose, therefore, that bee-keepers are right in removing queens after two years' laying: older queens with less vitality would produce colonies more apt to swarm, for such colonies would stand in greater need of renovation through new conditions in a new place. And it is quite natural that colonies with young, one-year-old, queens are less inclined to swarm, for the vital impulse in such colonies is strong. It is also natural that a cast is more mobile and flies farther: a colony sending off a second swarm must be in greater need of a change of conditions and the change must probably be more drastic. No wonder, then, that swarms arriving from considerable distances remain at an apiary for several years without attempting to swarm. And, finally, it is natural that apiaries moved into new localities, placed under new conditions, where the vitality of the bees is intensified, produce more honey in the first years after settling than the average yields at local apiaries.

The fact that a swarming colony leaves at home a *young*, newly-born, queen, furnishes extra proof of the correctness of our theory (although this is important for other reasons as well).

We think that the departure of the *old* queen with a swarm clinches the argument: the queen leaves the colony where during two or three years of her life she has laid 100,000 to 200,000 and more eggs.

In practice the occurrence of each swarm is always the result of an indirect influence of various external conditions, in the first place, probably, the state of the weather, the sources of food, etc. This circumstance has usually left in the shade the main cause—the degree of vitality, which, until recently, was not known and was, therefore, not taken into consideration. By drawing the attention of biologists to this aspect of living nature, Acadomician Lysenko has enabled them to turn a new leaf in the study of the nature of swarming. He has pointed out the spark that sets the colony a-fire making it swarm, he has indicated where the impulse for the sequel of events connected with swarming should be looked for.

Knowing the laws governing a phenomenon is as good as being able to direct it.

It is quite probable that by merely putting into a hive combs with sealed drone brood taken from another hive and thus making possible cross-fertilization of the virgins, the vitality of the colony may be secured and the cause of swarming eliminated.

When, in addition, the technique of controlling the mating of queens is well mastered and instrumental insemi- -nation becomes a simple process (we have had occasion to mention that the impossibility of controlling mating is

the main reason of lack of cultivated bee-breeds) bee-keepers will forget the anxieties of · natural swarming. Then
bee-colonies will cease to defy the will of man in casting
swarms which fly away from the apiary like fluff from a
dandelion flower.

THE PROLONGATION OF LIFE

What Determines the Difference in the Life Span of
Bees Bred in Different Seasons. "Summer" and "Win-
ter" Bees. Seasonal Variability in Bees and Protein
Content in Pollen. "Winter" Bees Bred in Summer.
A Part and the Whole.

It has been mentioned several times in this book that
bees emerged in spring or in summer live on the average
no more than six weeks and that those emerged in autumn
live six months and even more.

The fact that in autumn the colony is made up of hardy
bees, capable of living long enough and of passing through
the hardships of wintering without losing the ability to
nurse the first generation of spring bees, reveals one more
highly important feature of the biological intactness of
the bee-colony, of its harmony.

Thus we see that some generations of bees live six weeks
and others six, seven, and more, months.

Phenomena of this kind are described in science as "seasonal variability in the life span of individuals belonging
to different generations." But it is one thing to give a description of a natural law, and quite another to master it.

In a colony which has not superseded its queen, all the
bees of whatever generation are full sisters reared under
uniform conditions. They have the same parents, they have
developed from the same eggs in similar cells and have been
reared in the same way.

Should they not, then, have the same life span?

But we know that a queen emerges in the same colony
and from the same parents out of a similar egg, fed by the
same bees. She is capable of living five years, and we have
no data showing that a queen emerged in spring is shorter-

lived than one emerged in autumn. Consequently, the time of emergence may not be the decisive factor.

Then what is?

Just imagine three full sisters with the natural spans of life of six weeks, six months, and five years! The difference is simply staggering. For a being with a natural life span of six weeks, six months (to say nothing of sixty months) seem countless ages.

To have a clearer picture of these differences, let us represent them on another scale and consider life spans of 50 years, 200 years and 2,000 years.

Do not forget: in the three instances the embryos are absolutely alike—the pearly-white one-and-a-half-milli-metre-long eggs laid by the queen.

One cannot help being intrigued as to whence comes and by what is determined this astounding difference in the average life spans of creatures developed from embryos absolutely alike.

In the case of the queen the matter is simple enough: she, indeed, emerges from an egg in no way differing from those giving rise to worker-bees, but the grown queen larva is fed differently and is reared in a bigger and differently constructed cell. Besides, a queen leads a different life from that of the workers and we may suppose that these peculiarities of the rearing of the larva and the conditions of life of the perfect insect tend to increase its life span.

But workers of all generations are reared in quite the same way. Why, then, is a bee born in September capable of living at least five times as long as her full sister born in May?

Trying to find an answer to this question and recollecting some of the well-known facts of bee-colony biology, we discover at once a number of essential differences in the life stories of the May and September bees.

All "summer" bees (including our May bee) spend their lives in a growing colony, in a colony to which every day brings an increase of brood requiring food, in which combs are being built and a lot of energy spent on flights afield. The colony of that time is warmed by the sun through the hive walls and its food is nectar and pollen fresh from the blossoming trees and cereals.

"Winter" bees, born in September or later, spend their lives in a colony which does not grow, where the number of bees remains on the same level. Wintering, these bees stay within the hive huddled together in a cluster for several months, they do not build combs, they spend the stores collected in the summer.

Such are the differences in the lives of different bee generations which are apparent at first glance. But scientists wanted to know whether there are any inner differences in the nature of different generations of bees.

Only minute research could provide an answer to this question. That research was conducted by Anna Maurizio at the Liebefeldt Experimental Station in Switzerland.

Several hundred bees born on the same day in one and the same colony were put in cages, a hundred in each, where they received the same food. Temperature conditions were differ-

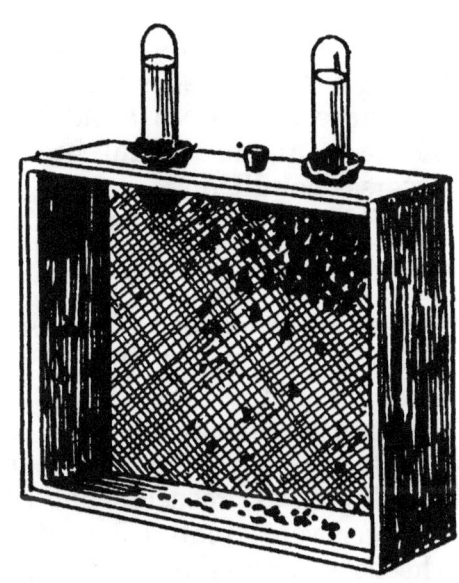

Cages with sliding front walls made of glass, in which bees are tested for longevity. The test-tubes in the ceiling of the cages are feeders containing syrup and water

ent in each cage. After several weeks of the experiment it became clear that at some temperatures the bees died sooner while at others they could live longer. There remained no doubt that the average life span of a bee depends on outward temperature.

But when, under another series of experiments, bees of the same age were kept under differing conditions of humidity, it became equally clear that the life span of a bee depends on the degree of humidity.

The experiments were carried further: groups of bees were kept under equal conditions of temperature, of humidity, etc., but were given different quantities of the same food, or equal quantities of different food, and it was established that the average life span of a bee is influenced by the quantity and quality of food consumed in the larval stage.

In addition, the nature of the bees' activities during the experiment was also found to influence the length of their lives.

We cannot describe all the variations of these painstakingly-conducted experiments: the more varied they were, the more data were accumulated, showing that the length of life of an individual bee is influenced by a multitude of diverse conditions.

This conclusion is incontestable, but, for all its correctness, it is too general to be of any practical use.

That is why, after all these experiments, more were carried on in order to find out whether there were any deeper-lying differences in the nature of different bee generations.

In the former series of experiments a comparison was made between bees of the same age kept under conditions similar in all respects but one. In the latter, uniform conditions were created for all the bees under experiment, but the bees themselves differed, although not in age but in the date of emergence.

Several hundred bees born on a May day in one colony were put into cages and placed in a thermostat with constant temperature and humidity. Two months later, bees born in that same colony on a July day were placed in the thermostat with the same conditions of temperature and humidity. Then the experiment was repeated with bees born in September. All the bees received food made to the same recipe.

A comparison of the respective lengths of life of the different bee generations under these artificially created uniform conditions was to show whether the conditions alone influence the life span of bees put into the thermostat immediately after emerging or whether these different generations differed in their inborn degree of vitality.

It became clear that bees born in May, July or September and immediately placed in the thermostat possessed different degrees of vitality. The average length of life of bees born in May and July was 600 hours, and of bees born in September, nearly 1,000 hours, i.e., almost double. The mathematical analysis of the data obtained as a result of hundreds of experiments has convincingly established this difference between the generations.

But in the present case it was necessary to probe to the bottom the cause of the greater longevity of "winter" bees.

Anatomists were the first to find this out: they proved that some internal organs and tissues of newly-born "summer" and "winter" bees were in different states. Thus, for instance, the adipose tissue and the brood-food glands are much better developed in "winter," longer-lived, bees.

Physiologists supplied additional evidence: the state of the adipose tissue and of the brood-food glands is determined by the protein content of food, i.e., the concentration of pollen in the larval food.

A by-product of the experiments was the conclusion that larvae fed on pollen collected by man produce somewhat shorter-lived bees than larvae fed on pollen collected from the same flowers by bees. This is one more proof of the correctness of the familiar conclusion from the experiments with cancer-infected mice that bees begin preparing protein food on the way from the flowers to the hive and manage, while flying, to change pollen considerably.

Biochemistry has long been aware that flowers of different plants blossoming at different times contain different quantities of digestible proteins.

At last there came a time when it became necessary to reduce to a system all the newly-acquired data.

The new data plainly showed that different quantities of pollen from different plants consumed by larvae influenced in different ways the development of the adipose tissue and of the glands in bees and thereby determined the possible length of their lives. It appeared that the presence and composition of food (and also the state of the colony and, first and foremost, the number of the larvae fed) determined whether the young bees would be short-lived, "summer" bees or "winter" bees capable of living much longer.

The most correct method of verifying this conclusion was to learn to obtain at will generations of "winter" bees in summer and of "summer" bees in winter.

It would be long and tedious to describe in detail the experiments whose main essence was changing, transform-

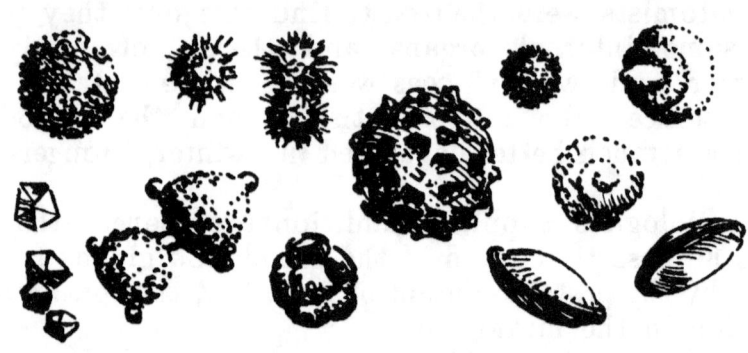

Pollen-grains of different plant species differ in form and size. Here pollen-grains are greatly enlarged

ing metabolism in the colony. We shall mention only the results of one such experiment: bees born in summer, i.e., those that should have lived not more than six or seven weeks, lived 100, 200, 300 and even 400 days.

The average longevity of a "summer" bee was multiplied by at least ten. The bees under the experiment lived ten average bee lives.

By postponing for such a long time the natural end, the experiment showed that the prolongation of an individual's life is a perfectly realizable task.

One must not suppose, however, that this was a case of "vanquishing" nature. Far from it. The phenomenon itself, the result of the experimenters' efforts, was not new in principle. This same phenomenon may be observed in bee life under natural conditions.

As a general rule, drones live about 100 days. But they die not because they cannot live longer: their life is forcibly ended by bees driving the drones from the hive in autumn. If the drones are left within, which usually happens in a queenless colony, they may live through the winter, altogether as long as 250 to 300, and more, days.

We think we have every reason to regard this phenomenon as one enabling the investigator of the bee-colony to peep into the innermost secrets of living nature.

Thus, as life shows, drones are capable of living much longer if they are not killed by bees, and experiment shows that worker-bees are also capable of living longer if they are not killed by conditions.

We may well suppose that the life-impulse in the embryos is sufficiently strong to enable the insects developing from them to live much longer—in the given instances at least five to ten times as long as is actually the case.

The longevity of a bee largely depends on the age and condition of the bees that fed it as a larva

Such being the fact, the coming of the end can be postponed; and basing himself on the vital sources possessed by organisms, man can prolong life far beyond its accustomed limits.

This conclusion is important from many points of view and, we cannot help noting here, it is successfully proved by the practice of prolonging the life of whole bee-colonies.

A bee-colony is born at the moment when it leaves its old hive in the shape of a swarm. This marks the beginning of a colony's individual life. Under favourable conditions, if all goes well, such a colony divides by casting a swarm in a year or two. After this, the new colony with a vigorous young queen lives, develops and undergoes the process of rejuvenation while the swarm with the old queen is rejuvenated by changing the old conditions for new ones and, building new combs, increases with redoubled vigour.

No matter how many times swarming occurs, the colony living in the old hive finally perishes due to some cause or other.

In his *Dialectics of Nature* Engels dwells at length on the problem of life and death. Death must be regarded as an essential moment of life; actually, life itself contains the negation of life; life is always pictured in connection with its necessary result—death, the embryo of which is always contained in life. This is the essence of the dialectical understanding of the life of an organism, Engels said, and he concluded with the words: "Living means dying."

The mortality of organisms is at the same time an adaptation for the survival of the species. An organism can live only in unity with the conditions that have bred it. In the course of long periods of time, unavoidable changes of climate and other external conditions would unavoidably break this unity and eliminate biological species. The preservation, life, development and perfection of biological species is possible only because of the succession of generations of living organisms.

This is what materialism means by the life of a species. Dying means living.

Consequently, the honey-bee lives as a biological species owing to the colony being mortal like an individual organism.

The death of bee-colonies may be often observed not only under natural conditions but also under the conditions of skep bee-keeping, very close to the natural ones.

The description of the death of a bee-colony in Leo Tolstoy's *War and Peace*, Part Three, is well familiar to many. The great writer gives an unsurpassed picture of the dying colony, impressing the reader's senses of vision, hearing, smell, and even touch.

"There is not the fragrance, the hum of life. A tap on the hive does not produce the general and immediate revolt of thousands of little creatures, curling themselves round to sting, buzzing and fluttering with rage, and filling the air with the stir of busy labour, though here and there, in its depths a feeble hum may be heard. At the entrance there is no heavy, aromatic scent of honey, no warm odour of gathered stores. No watchful guards are there, ready to give a trumpet call of warning, and then to sacrifice their lives in defending the commonwealth. There is no peaceful, regular toil betraying itself in a steady murmur; only a fitful and broken buzz. . . . Instead of swarming bunches of honey-laden bees, clinging to each other, or brushing off the pellets of gathered wax, only a few torpid and half-dead insects are to be seen at the bottom of the hive, or wandering idly and vaguely about the fragile partitions. Where once there was a smooth floor, clean-swept by the fanning of their wings, the seams neatly caulked with wax, lie scattered crumbs of wax, broken ruins, a few dying creatures with legs still quivering, or corpses left unburied. . . . .

"Where, not long since, thousands of bees stood in circles, and back to back, watching the mysteries of hatching broods, there is only a sprinkling of exhausted workers. . . . It is a realm of death and decay! The few that survive climb, try to fly, cling to the masters' hand, and are too weak even to sting ere they die."

On discovering through the peep-hole of a log hive a colony in such a state, a bee-keeper marked the hive and on his first free day removed the combs to be melted for beeswax.

All this was almost unavoidable as long as bees were kept in skeps and log hives which did not allow the bee-keeper to look well inside. Since bees have been housed in movable-frame hives which can be thoroughly inspected and the frames in them easily exchanged, the death of bee-colonies can be prevented.

Today the death of a colony is usually the result of an accident, of an oversight or a mistake of the bee-keeper's. No sooner are alarming symptoms such as the loss of the queen or the absence of brood noticed in a hive, than the bee-keeper can put into it a piece of comb with young larvae from a prosperous colony and, if the colony has not gone far along the road of destruction, this simple operation, in many respects similar to the grafting of tissues, prevents the colony's death. The colony rears a new queen from the foreign brood and recovering in a short time, resumes the normal course of its life.

From time immemorial bee-keepers have used this method of saving colonies, owing to which the life-process is prolonged indefinitely. Colonies are known to have been preserved by this means for 100, 150 and 200 years. With bees, such prolongation of life is the more easily accomplished as the place of old bees, worn-out and gradually dying away, is taken by young ones.

The succession of generations imperceptibly taking one another's place makes the colony an organic whole, constantly self-rejuvenated, regularly self-renovated and capable of preserving vitality and viability as long as a skilful apiarist protects it from death.

All living nature shows that the first line of its development star s from amorphous protein to the cell and from the cell to the multicellular organism. With this line is fused the other, developing from the individual (unicellular and multicellular) to the colony, from the ever-growing colony, perfecting its organization to an organized colony, an organic community in which the very concept of the individual has expanded and become relative, and at a certain stage of the development of which, as Engels said, "the concept of the individual cannot be established at all sharply. Not only as to whether a particular animal is an individual or a colony, but also where in development one individual ceases and the other begins."

DIRECTED FLIGHT

CONTROLLED HIVE

End of the Bee-Killing Practice. Can Bee-Keeping Be a Profession? The Bee-Entrance and the Wind. Stages in the Development of the Theory of Bee-Keeping. One Must Know the Laws of Bee Life in Order to Control the Behaviour of Bees.

Bee-keeping is justly considered one of the most ancient trades, but whereas cultural practices and selective breeding were long ago used in horse-breeding, animal raising, poultry breeding, fruit, vegetable and grain growing, a predatory system was still in vogue in bee-keeping, under which bees were killed by sulphur fumes at the end of the summer and the honey taken.

The foundations of rational bee-keeping were laid not more than 150 years ago by Prokopovich (usually spelt Prokopovitsch) and Huber who lived and worked early in the nineteenth century.

François Huber, a blind Swiss naturalist, and his loyal assistant François Burnens, discovered many important facts in the biology of the bee-colony.

In 1828, Pyotr Ivanovich Prokopovich founded in Chernigov Gubernia the first bee-keeping school in the history of apiculture. His experimental apiary grew to number about 10,000 colonies and was one of the biggest in the world. The method of bee-keeping he worked out put an end to the killing of bees, widely practised at the time, while the hanging-frame hive and the queen-excluder invented by him enabled apiarists to control the life of bees and for the first time to obtain pure honey in supers. Very big crops of honey were taken from such record colonies in Prokopovich's apiary as *Arkhangelsk* and *Siam*. Prokopovich was perfectly right in saying: "I have penetrated deeper than any of my predecessors into the mysteries of the bee world."

It was this deep penetration into bee biology that enabled him to elaborate his "science of bees along such lines as to make it possible, without killing them and even through maintaining their existence by various sure methods, to preserve each bee-colony for ever, or, in other words, to keep bees *permanently* in a hive once a colony has been put there."

In Prokopovich's lifetime his contemporaries owned that his method "opens perfectly new roads before bee-keeping and raises it to the dignity of a science. Nothing here is left to chance, to luck, and such-like, but everything has its cause and the consequences and conclusions thereof are confirmed in practice."

It would not be too much to say that bee-keeping on a large scale began with Prokopovich.

Soviet people cannot regard the economics of bee-keeping in capitalist countries as anything but a string of absurdities. Strange as it seems to us, years of plentiful honey-yields impede the development of apiculture. Equally strange, but none the less true, is the fact that the specialization of bee-keepers, their growing mastery over bees also serve as brakes to the development of the trade. The fact that bees produce honey ready for use, too, is against the bee-keepers' interests.

It has been announced that a serious drawback of American apiculture lies in food-manufacturers not being interested in honey, because there is not much they can do with it. "Natural honey needs no processing." The modicum of truth in this statement is, probably, that trading in a product that requires no human labour for its production does not pay. Bees do not produce surplus value. All this may seem incredible, but such are the ideas, however incompatible with common sense, that are being propounded in books and articles by leading bourgeois apiarists.

In works supplemented with statistical data and computations they prove that the drop in the prices of honey due to abundant honey crops results in ruining the apiarists and in reducing the number of apiaries so that after rich yields the number of bees 'in a country decreases sharply and part of the crops remain unpollinated.

The American agricultural press, and fruit-growers' magazines in particular, can no longer keep silent about the lack of pollinating insects, which in some localities has become disastrous. Certain enterprising persons seized this opportunity of advertizing the "pollen bomb" which, they claimed, was capable of doing the work of the honey-bee. This bomb, of course, in no way helped fruit-growers but caused great prejudice to apiarists and was profitable only to those who made and sold it.

The very creation of such a bomb is illustrative of the relations existing between apiarists and fruit-growers in the U.S.A.

An American economist held that farmers renting land to apiarists must pay the apiarists for the pollinating work their bees performed. Then farmers countered with the claim that they should be paid not only for the land but also for the nectar collected from their flowers. The argument as to whether apiarists should pay for nectar or farmers for pollinating services went on for many years.

It should be recalled that both in the U.S.A. and in the capitalist countries of Europe, as in prerevolutionary Russia, the average annual increase of bee-colonies has not exceeded 3 to 4 per cent for the last decades. On the other hand, the annual increase of bee-colonies in the collective-farm apiaries in the U.S.S.R. during the ten years preceding the Great Patriotic War was 14%, i. e., almost four times more. Thus we see that collective-farm apiarists were not discouraged by the fact that bees produce honey ready for use.

In the U.S.S.R. the bee-keeping industry has been placed on a firm foundation for the first time in history; the apiary is a planned branch of collective-farm and state-farm economy and bee-keepers have at last been enabled to discard amateur practices and devote themselves heart and soul to bee-keeping as a profession. It is absolutely essential to put one's heart into this difficult work many aspects of which are yet unknown and in which many important phenomena still are beyond the control of man.

Spring and summer differ from one locality to another and in the same locality they may differ from year to year. Here weather is a decisive factor.

But, like the captain of a sailing ship, whose duty it is to bring his craft safely to port, adverse winds and dead calm notwithstanding, the collective-farm bee-keeper must produce a planned quantity of honey and wax and see that the crops are pollinated irrespective of weather and other conditions. And the bee-keeper has to foresee many things when, in spring, he takes the bees out from their winter quarters. Hives with colonies that increase slowly and take a long time to build up to strength should be placed so that the first rays should fall on them and the sun should warm them as long as possible; other hives, where the colonies are inclined to swarm, should, on the contrary, be placed in cooler spots where the shade of a tree may protect them from the sun at the hottest hours of the day. Both kinds of colony should be kept well protected from winds.

The better a bee-keeper is able to think of such and similar details of his work, the more obedient his bees will be. He must increase their number in each hive early in spring so as to be able to send them, like arrows from a bow, to the flowers whenever the honey-flow sets in.

The legions of bees that will fly afield at the blooming time of the major honey-plants emerge and mature long before the time comes.

While the first shoots of the future honey-plants turn green, before some of the plants are even sown, before the flower-buds on the trees begin to swell, the bee-keeper has done all that is necessary for thousands and thousands of future foragers to grow in their cells, to pass through the larval stage, to moult and become pupae. . . . He directs the activities of his bees so that by the time of the honey-flow there should be less work to do within the hive, especially work connected with brood-rearing.

The new bees must bite through the wax cappings of their cradle-cells long before the main honey-flow sets in so as to pass through the sequence of home duties by the time the first joy dances of the scouts' announce the presence of nectar in the flowers and the young bees catch the foraging fever. From that moment on the entire colony will concentrate all its energies on collecting and storing nectar.

This is the course of events at an apiary where the bee-keeper does not follow a beaten path in the management of his bees but thoughtfully watches the progress of each spring and summer, taking heed of all peculiarities of the seasons. Each peculiarity he has been able to assess correctly influences his plan, according to which the growing foragers are to work.

The skill, even art, of a bee-keeper consists in being able to time the population increase of the colonies to the blossoming of flowers, to utilize all the energy of the colonies for filling the combs with honey. With this aim in view, a skilful and persistent bee-keeper manages to bend to his will the biological processes in the bee-colony. He uses various methods to stimulate or hold up egg-laying, to increase or decrease population, to balance the number of bees of various ages to suit his purpose. With perfect assurance he divides a colony so that the bees will build emergency queen cells. Sometimes he unites two colonies into one. He breeds drones or prevents their appearance. He feeds the bees or takes away honey-filled combs. He makes bees secrete more wax, draw comb foundation, renovate old combs or, on the contrary, prevents them from engaging in building activities. He does not take his eyes off the bees for a single day and directs their life through the laws governing the bee-colony.

Bees obey an experienced and skilful apiarist wherever he may keep them: in a collective-farm apiary, in a collective farmer's cherry orchard, in a worker's front garden in the Urals, on a seventh-storey balcony in Polyanka Street, Moscow (keen apiarists keep bees even in cities), on the window-sill in the study of a naturalist who spares now and then a few minutes from his work to attend to his bees in a hive with the lobby leading from the entrance, through the window-frame into the garden, and even on house-roofs where doves used to coo.

BEES UNDER GLASS AND GAUZE NETS

Beehives in Green-Houses. Willow Blossoming under
Glass. The Task I. V. Michurin Set the Bee-keeper
I. A. Kiryukhin. A Pollinating Instrument under a Gauze
Net. The Nucleus.

Honey ceased long ago to be the only source of sweet
food for man. Through selection sugar-beet has been made
sweeter than nectar, and every year sugar refineries working sugar-beet crops produce mountains of dry, imperish-
able and cheap sugar whose transportation presents no
difficulty whatever. But this has in no way lessened the
demand for natural honey; on the contrary, it is growing
constantly.

Beeswax that used to burn in the shape of candles has
long been used in the electrical-equipment industry. The
substance which, if the legend is true, was employed by
the pioneers of aeronautics for making an artificial wing
is used today in aviation. Beeswax is required in metal-
lurgy—for high-quality castings, in railway transport—
as a component part in mixtures for brakes, in factories
producing optical instruments—in glass engraving. The motor car, the polygraphic, the glass, the varnish-and-colour,
and many other industries have various uses for wax.

Bees, too, are being used for new purposes. Thus, plans
provide for moving bees into orchards for pollinating the
blossoming fruit-trees and berry-bushes, which is of no
less importance for the crops than introducing fertilizers
into the soil.

Today bees are needed in hot-houses and wherever vegetables are grown under glass, for, in defiance of the old
saying that "Every vegetable arrives at its own season,"
vegetable-growers now follow the motto: "Every vegetable
in every season."

Nowadays, various vegetable plants, bushes and even
fruit-trees grow, blossom and bear fruit under transparent
glass roofs, within glass walls, where the air is heated by
cleverly-constructed stoves and hot-water pipes and where
electric bulbs automatically moving at a certain height
from one end to the other provide extra light to young
shoots, seedlings and grown plants. Our towns are now girded

by hot-houses and in the shops one may often see green cucumbers in December or fresh. strawberries in February. To obtain a crop of cucumbers under glass, cucumber flowers must be artificially pollinated with a brush. This is a laborious process requiring jeweller-like precision; the work proceeds slowly and results in a large percentage of crippled fruit—mis-shapen, undersized and meagre. But blossoms pollinated by bees set good fruit.

Then a hive was brought into an empty hot-house in winter.

Warmed by the inside air, bees left the cluster and took a cleansing flight. A day or two later the hive was put in another hot-house where plants were in bloom. The colony started its activities although only a thin layer of glass protected it from the outside frosts.

The pollination of the blossoms proceeded quite satisfactorily.

True, bees cannot collect much nectar under glass, but a poor honey-flow can be easily supplemented by combs of honey or syrup feed given the bees. Then there is not enough pollen and if the brood does not receive enough food, for which pollen is necessary, its development will cease and it will eventually perish. The colony will cease to grow and will become weaker. This may be prevented by bringing branches of willow with flower-buds into the hot-house. A few days after the branches have been put in water (even if they were cut off the trees during severe frosts) the buds on them will burst and male flowers appear. From these the bees will collect fresh pollen which can be used in the production of brood food.

On one hectare under glass, bees do work equivalent to from 2,000 to 2,500 man-days; the crop of cucumbers doubles, as was seen in the collective-farm hot-houses in Moscow Region, and the quality of fruit largely improves, the percentage of "cripples" being negligible.

A medium colony efficiently services a hot-house with an area of 3,000 square metres at the First Vegetable Factory, in Moscow, with the result that about 10 workers are freed from tedious work, considerable sums are saved and the yields increase.

Need we speak of the importance of bees in the pollination of plants grown in hot-houses beyond the Polar Circle?

It should be noted that bees treat many plants under glass in a different way from those in the field.

In the field the flowers of kok-saghyz, for instance, are visited in order to collect nectar, but in a green-house, where this plant blossoms in early spring, foragers also gather pollen from it.

Bees are made to work not only in hot-houses; they perform even more important functions under gauze nets where fruit-growers place them.

When a selectionist wants to pollinate a variety artificially with the pollen of some other variety or with a pollen-mixture, the scope of his work is limited by the blossoming periods of the plants and by the number of his assistants. Under the best of conditions he is practically unable to pollinate all the flowers of a tree of any size, for there may be thousands of flowers on it. And he pollinates by hand, although he is perfectly aware of the shortcomings of enforced pollination. I. V. Michurin thought of enrolling the assistance of bees. He charged I. A. Kiryukhin, a pupil and collaborator of his, to work out new methods of crossing.

To ensure artificial pollination with indisputably pure pollen, Kiryukhin used only young bees that had not left hive before, so that there were no pollen grains on their bodies.

Combs with 3,000 to 4,000 young bees are put into a hive with an entrance through which bees can enter the hive but cannot go out, and an exit, also allowing only one way passage. When a bee leaves the hive through the exit, she enters a porcelain tube containing fresh pollen of the needed variety. While she passes along this tube, the bee gets covered all over with the pollen.

This pollinating instrument containing several thousand bees is put under a gauze net covering the tree to be pollinated. The net is spread long before blossoming time to prevent other insects from entering and the bees from escaping.

The nucleus is put first near the tree outside the net and the entrance and exit are opened. Old foragers take the opportunity of flying home and only young bees remain within. After this division, the nucleus is put under the gauze net and fresh pollen is added in the porcelain tube from time to time.

Bees leave the hive covered with pollen, fly to the tree and, visiting one flower after another, pollinate them with the chosen variety. While the most experienced and skilful worker can hand-pollinate not more than 600 flowers during an eight-hour working day, bees visit as many as a quarter of a million flowers in that time.

The knowledge of the laws of biology enables man to utilize them for his purposes and to exercise control over bees in the hive.

But man has long stood in need of controlling the activities of bees outside the hive. There, in the open air, bees were beyond his control: they left the hive in response to the dancers' call and flew to the address these had communicated, and man was at a loss how he could direct their flight. Bees working under glass or gauze nets was the beginning of man's control outside the hive walls but it was not yet work in the open.

With a gauze net you can cover a few flowers, one tree, two, five or even ten trees. But you cannot cover an orchard or a field, which is exactly what plant-growers need. They want to guarantee the pollination of large plantations and crops. In particular they now want bees to pollinate alfalfa and red clover.

BEES AND THE IMPROVEMENT OF CULTURAL PRACTICES IN SOCIALIST AGRICULTURE

Laws Known Are Laws Mastered. Grass Restores Fatigued Soils. Why Clover Yields Little Seed. The Bumble-Bee as a Pest and the Bumble-Bee as a Pollinator.

The improvement of cultural practices in socialist agriculture was started long ago and each five-year plan period brings about an intensification of this work.

Crop- and fodder-rotation is spreading over greater and greater acreages of collective-farm and state-farm cultivated areas. Land is being utilized ever more profitably. Owing to up-to-date equipment being supplanted by up-to-the-minute equipment, land cultivation is improving apace. Mineral fertilizers are being perfected and used on an increasing scale, while agronomists are constantly improving methods of applying them. Varieties with good crop capacity are superseded by those with still better crop capacity and more suited to local conditions.

Collective farmers and workers of state farms regulate the water-supply and the feeding of the crops, thereby making the land a permanent means of production which, far from deteriorating is perfected through constant use. Correct crop-rotation plays an important part in this process.

In those far-off times when agronomy was not so fully aware of the importance of leguminous plants as it is now, when the action of these plants sown together with cereal crops was not yet fully clear, Kliment Timiryazev stated that the introduction of leguminous crops, in the first place, clover, into crop-rotation was a remarkable discovery highly beneficial to mankind.

Today the importance of this discovery has become still more patent.

The significance of this discovery lies, as we know, not only in the fact that clover or alfalfa are capable of yielding in many localities great amounts of cheap green fodder and hay.

Practical agriculture knows fodder crops possessing higher productivity even than these, as for instance, sugar-beet in which both the root and the leaves can be utilized or maize yielding grain, green fodder, and—as silo—carbo-hydrate fodder (stalks and leaves) and protein concentrates (grain in the early stage of wax ripeness ensiled together with the cobs). As a matter of fact, maize can be siloed even at an earlier stage—the stage of milk ripeness—which allows this crop to be cultivated in a number of areas in the North and East of the U.S.S.R. where formerly, due to the considerable length of its vegetation period, the cultivation of maize was considered impossible Today in many of these regions the value of maize as silo crop is higher than that of fodder grasses.

What, then, is the significance of fodder grasses? Does it lie in the fact that clover and alfalfa can accumulate in the soil great quantities of the most important plant food—nitrogen—owing to the presence in their tubers of nodular bacteria? Under red clover, for instance, the soil accumulates nearly two centners of nitrogen per hectare, the equivalent of forty tons of manure.

Although this property of fodder grasses is of much importance, it does not, however, constitute the decisive factor: agriculture in the U.S.S.R. has vast sources of nitrogen at its disposal. Each succeeding five-year plan brings about an ever greater development of animal husbandry which will yield a considerable quantity of nitrogen in manure. The country has a large-scale chemical fertilizer industry capable of increasing from year to year the output of nitrogen to satisfy the needs of agriculture. At the same time Soviet agricultural microbiology has scored great achievements in the production of bacterial fertilizers, such as *nitrobacterin, azotogen,* and others. The technology of composting fertilizers has made big strides.

The most valuable property of clover, alfalfa, and other perennial legumes is that, sown together with cereal crops in localities where they yield abundant crops of green mass above ground, they yield equally big crops of tubers underground. The physical, chemical and biological consequences of growing perennial grasses in a field are that, since the field is not ploughed, the soil structure is reconstructed and the soil quickly becomes fertile again.

A soil exhausted by ploughing, if left fallow, is reconstructed but slowly: nature has no cause for hurry and it takes some 15 to 20 *years* to complete the process of reconstruction. But when sown with a mixture of grasses including clover and alfalfa, the soil is reconstructed in a *year or two.* These leguminous plants are capable of quickly restoring the soil's structure, which is the prime prerequisite of fertility.

A truly scientific, creative application of agronomic theory, free from dogmatism, leads to progress in enriching soils.

In exposing the short-sighted and predatory character of agriculture under private landowners, Marx in his *Capital* said: "A whole society, a nation, or even all societies together, are not the *owners* of the globe. They are only its *possessors*, its users, and they have to hand it down to the coming generations in an improved condition, like good fathers of families."

This idea of the founder of scientific communism is being realized. By making land the property of the entire country and organizing socialist, collective-farm economy, the Soviet people have turned a new page in the history of agriculture.

Every year sees them introducing new improvements in agriculture and, like good fathers of families, they will hand down to the coming generations a better organized, more fertile land.

Agronomists come to a clover field and count the number of corollas per square metre of control plots. There may be as many as 500, each consisting of about 100 pink florets, which makes 50,000 florets a square metre and 500,000,000 a hectare. When there is a good crop of clover there may be 1,500 million florets and more per hectare.

Agronomists enter these figures in their diaries and come to the control plots about two months later. The plants are drying and the corollas have turned brown. Computations show the number of set seeds. A thousand clover grains weigh about 1.5 grammes; the 500,000,000 florets on a hectare must yield from 6 to 7 centners of seed. Actually, however, the

Sainfoin has long been known as a fodder crop and an excellent honey-plant

crop gathered from a hectare amounts to no more than 60 or 70 kilogrammes.

We have mentioned above that clover is self-sterile and cannot be pollinated either by the pollen of the same flower or by that of other flowers on the same plant. Seed sets only as the result of pollination with the pollen from another plant. The best set of seeds is obtained in the case when the pistil receives foreign pollen at least twice or three times.

How can this be done with each of the 500,000,000 or 1,000 million florets that computations show to be growing on a hectare? And what about hundreds of hectares? This enormous volume of work is absolutely beyond human power and is done by insects visiting clover to collect pollen and nectar.

Bumble-bees pollinate red clover well. True, some bumble-bees, notably Bombus lecorium do not give themselves the trouble to suck the nectar from the nectaries deep in the florets but bite through the lower part of the corolla and suck the nectar with their proboscides through the hole. The flower thus robbed of its nectar remains unpollinated. But other bumble-bee species extract the nectar in a legitimate way—through the upper entrance of the corolla—and in so doing pollinate the flower.

But the trouble is that there are few bumble-bees left in nature today. Bumble-bees build their nests in virgin soil or soil that has not been ploughed for a considerable time, and as such land is being cultivated on an increasing scale, there are fewer bumble-bees every year. In addition, the number of bumble-bees is influenced by various meteorological causes and entomologists know of favourable and unfavourable years for bumble-bees. All the same, even in the best of years, there are never enough bumble-bees to pollinate all the flowers in the ever-expanding fields of clover which are capable of feeding a much greater number of nectar-gatherers. And agronomists are perfectly justified in suggesting that bumble-bees should be bred specially for pollinating red clover.

It is pretty much the same with alfalfa which often yields poor crops of seed owing to lack of insect-pollinators capable of opening or "tripping" and properly pollinating its peculiarly-constructed flowers.

As Academician Lysenko says: "Clover and alfalfa seed crops depend in the main on the extent to which the seed plots are provided with pollinator insects and protected against attacks by pests. If practical agriculturists do not interfere with this process of ordinary biological life, do not promote the development of useful insects, do not prevent the development of pests or destroy them, it will, as a rule, be impossible to get a sufficiency of seeds, not to speak of a surplus of them."

It is a known fact that in the chain of mutual connections the complex and the simple, the great and the small are interdependent, and we see that the planned intensification of agriculture over the vast expanses of the Soviet Union is dependent in one of its important aspects on a tiny insect whose business it is to reach its tongue down the corolla of red clover or under the yellow or blue vane of alfalfa flower. Stable and abundant red clover and alfalfa seed crops are necessary for the introduction of correct croprotation.

Nothing will prevent the Soviet people from accomplishing their task of the refashioning of the earth for obtaining stable and high yields and increasing the fertility of soils, to which the peoples of the U.S.S.R. have devoted themselves with such enthusiasm. Soviet agronomy has found ways and means of solving the problem of clover and alfalfa pollination.

Blue sown alfalfa, as well as yellow and hybrid alfalfa is best pollinated by some solitary bee-species

BEES ON RED CLOVER

Bees' Proboscides under the Microscope. A Treasure
of Nectar at the Bottom of a Flower. New Studies of
the Instincts and Reflexes of Bees. From Heather to
Clover and Back. First Successes in the Training of
Bees.

The list of insect-pollinated agricultural crops in the
U.S.S.R. exceeds 100 varieties and areas sown to these
crops amount to 15 per cent of all cultivated lands in the
Soviet Union. Bees make up 75 per cent of all the insects
visiting the blossoms of the agricultural crops. As has been
mentioned above, the value of crop yields obtained through
the pollinating services of bees is from 10 to 15 times as
great as the value of honey and wax produced by bees in
the best of years.

All these facts have long been known, and when "from
cell of wax does fly a bee to gather in sweet contribution,"
as the poet says, agronomists cannot be indifferent as to
the source of this contribution, whether honey is collected
from weeds or from cultivated plants.

This is the more important as there is such a large
percentage of sterile red-clover blossoms, hundreds of millions of red florets in corollas left unpollinated.

Professor A. F. Gubin, son and grandson of bee-keepers
and pupil of V. R. Williams, said that although there was
a dearth of bumble-bees we could keep as many honey-
bees as were necessary. If bees were capable of pollinating
red clover they would directly increase the seed crop from
the 15 per cent of cultivated areas under insect-pollinated
crops and indirectly—through clover in grain mixtures—
would increase the yields of all cultures in crop-rotation,
bringing about a general improvement in agriculture.

In their books, naturalists, entomologists, botanists
and zoologists, held that bees never visit red clover, because, the average length of their proboscides being six
millimetres, they cannot reach the nectar at the bottom
of the clover floret which is almost ten millimetres deep.
Later on it was established that a drop of nectar at the
bottom of a floret can rise as in a capillary tube between

the inner wall of the floret and the style of the pistil or a filament. Thanks to this phenomenon, the useful length of a bee's tongue can often be considered to be more than its anatomic length. The four millimetre gap between the flabellum on the end of the proboscis and the surface of the nectar drop in the floret is shortened, but alas, not

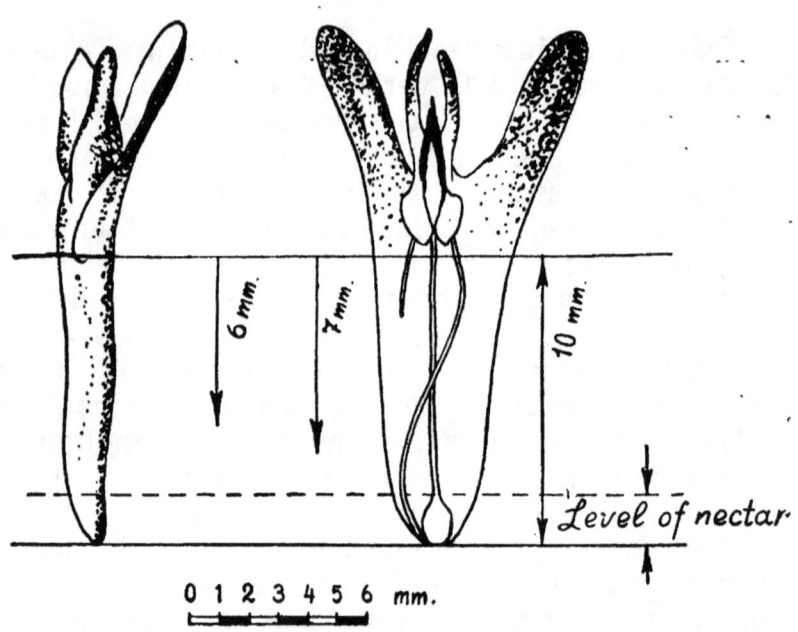

A diagram of a floret of red clover. *Left*: natural appearance. *Right*: corolla given in a section, ovary, pistil and accrete stamens being visible

enough for the bee to reach the nectar. This was advanced as the explanation why bees do not visit clover although its flowers contain a considerable quantity of nectar and pollen.

Then, on the initiative of Academician N. M. Kulagin, a specialist in apiculture, entomologists started studying bee tongues with the help of micrometers. A few years later agronomists and apiculturists throughout the world were startled by the announcement of the Russian agronomist I. N. Klingen who, studying the tongue lengths of various races and strains of bees, had established that the average tongue length of the Caucasian bees is a millimetre greater than that of the honey-bee of Central Russia.

At last there came a ray of hope that clover crops would cease to be dependent on the pollinating services of bumble-bees. Caucasian bees became widely popular and were dubbed "red-clover" bees. Endless praises were sung to them at agronomic congresses and in agronomic papers and magazines.

This started the importation of Caucasians into northern regions where clover is cultivated on a large scale. But the state of things remained almost unchanged.

Agrotechnicians and plant-growers have been engaged to this day in inventing such methods of cultivating and fertilizing land under clover as would cause a more intense secretion of nectar. If the level of nectar in the flowers were higher, bees would willingly visit red clover.

And today selectionist plant-growers dream of clover with shorter floret tubes and selectionist-apiarists cherish equally sweet dreams of bees with longer tongues.

And although no such clover and no such bees exist in nature, many an apiarist or agronomist has seen bees on red clover.

Both the corolla and the florets of clover are very much unlike the blossoms of other leguminous plants. The vane and the keel usually seen in leguminous plants are changed almost beyond recognition in the clover, but still they are to be found there, too.

Professor B. N. Shvanvich of Leningrad observed that introducing her proboscis into a floret, a bee moves the vane with her head and opens sidewise the halves of the keel. The stamens growing together and forming a style which is pressed against the folded ventral petals of the keel, slip from their place and press from below against the bee's mandibles while she is working her tongue within the floret tube. Pollen sticks to the ventral side of the proboscis and to the mandible; from here the bee removes it with the spikes on her legs. The stigma which extrudes together with the stamens touches the bee's mandible and becomes covered with pollen mixture.

But all those details became known at a later date.

Professor A. F. Gubin, too, had numerous occasions of observing short-tongued common bees of Central Russia on red clover. He became interested in the phenomenon and had observations made on a large scale.

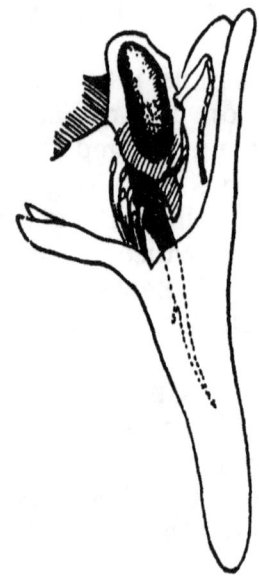

These observations brought quite unexpected results.

It was proved that forest and steppe honey-bees with short tongue lengths can be found on red clover in as great (it would be more correct to say: in as small) a number as bees with longer tongues: only one or two bees out of every hundred visit red clover. And even these may be scouts looking for honey-plants.

The data further showed that the farther was a field situated from an apiary the fewer bees there were on it: each hundred metres between the field and the apiary brought a 4 per cent decrease in the number of bees in the field. This meant in practice that with an apiary situated two or three kilometres from a clover field no bees visited it.

This would be a minor ill, for an apiary could be moved right into the field and the hives put at both ends, so that bees might fly in opposite directions and become scattered all over the field.

When a forager thrusts her head into a tube of a red-clover floret, the stigma and anthers of the flowers press against the bee's mentum from below, covering it with grains of sticky pollen which is thus carried from plant to plant

Computations were made concerning the rate at which bees work flowers, the number of foragers in a colony, their flight ranges, the average number of flowers per hectare and the average blossoming period of clover flowers. These computations led to the following conclusion: with one or two per cent of foragers visiting red clover, an apiary of from 60 to 120 colonies may be enough for pollinating clover over a sizable plot.

Practice corroborated the data of the computations. Several hectares of clover in the vicinity of big apiaries yielded from three to five, and more, centners of clover seeds.

So it was clear that Professor Gubin's undertaking was not hopeless, provided collective farms had more apiaries, more hives in the apiaries, and more bees in the hives.

The task could be solved with smaller apiaries if more than just two per cent of the bees could be induced to visit clover. But how could this be done?

A simple answer suggested itself to this question: By removing plants competing with clover. Indeed, if by the blossoming-time of clover hives were placed properly, other honey-plants mowed within the flight range of the bees, the blossoming-time of lindens postponed to a later date (this can be done by piling large mounds of snow around the trees in spring), would not the bees be compelled to visit clover from which they are usually kept away by other plants?

In spite of all the difficulties in the way, this new experiment was carried out. At last one would know for certain whether bees can be made to work clover if deprived of all other sources of nectar. The answer was a disheartening one: the fewer other plants grew around the apiary, the fewer bees visited clover. With a poor honey-flow the bees slackened their work, many of them remaining idle in the hive.

More than that: it was established that when favourite honey-plants grew near red clover and blossomed at the same time as it, the number of bees on the clover increased. True, the increase was but an inconsiderable one; out of a thousand bees 30 instead of 20 visited clover. Still this made one think that by adding white clover, buckwheat, mustard or melilot to red clover the pollination of a seed plot might be ensured by 50 to 100 colonies instead of 60 to 120.

But then not every locality has apiaries even as big as ·that.

Some other way of solving the task had to be found. And it is often useful in solving big problems to be able to start with little things. In analysing one must not lose sight of synthesis, i.e., in solving particular problems one must not forget the goal, which must always be practical results and not studying reality for the sake of study.

To start with, a simple experiment was carried out.

A feeder containing mint syrup was put into a hive over the frames. Two pieces of filter paper were put under the landing-board: one soaked in pure water and the other in mint essence. On leaving the hive the bees, accustomed as they were to the smell of mint, settled on the paper smelling of mint. The answer was clear: the fragrant beacon had served its purpose.

Then the bees were set to solve a more difficult problem.

The mint-syrup feeder was left in the hive and at some distance in front of the hive were placed four saucers: one of them was filled with pure water, another contained pure sugar syrup, the third mint water, and the fourth mint syrup.

Bees from all other hives passed by the saucers while bees from the hive under experiment were interested in their contents. But not all the saucers proved equally attractive to the bees. None of the bees alighted on the saucer of water, 23 alighted on the one with sugar syrup, 62 on the one with mint water and 131 on the saucer with mint syrup.

The next day no feeder was put into the hive; the experiment was repeated with similar results. The bees seemed to say: "We remember the smell of yesterday's feed!" The experiment was repeated twice, also without putting the feeder into the hive, and the number of bees on the saucers quickly diminished. In this way the length of bees' "smell memory" was established.

Perhaps the smell of mint oil is some exception? So feeders containing caraway, lily-of-the-valley, white-clover and colza syrups were put into hives in a new series of experiments and the number of bees showed that they remember different smells equally well and obey the will of the bee-keepers as expressed by the syrups in the feeders.

The last experiment was conducted with lilac syrup. The action of the syrup was controlled not by saucers containing syrup but by bushes of blossoming lilac. On that day the bees worked lilac intensively until it became dark.

Perhaps this showed the way to direct bees to red clover. But the first attempts to utilize this method in the case of red clover proved a failure.

True, after being fed clover-scented syrup the percentage of bees on clover doubled, a goodly increase, as far as figures go. But actually it was two or three bees instead of one or two out of a hundred. It looked as if the bees obeyed the command to go afield, but once there, refused to do what they were told to.

Some time passed before it was discovered that bees sent afield by the smell of the feeder were "seduced" by the competing plants whose removal had been so ineffective in previous experiments. Observations showed that other honey-plants from which bees had collected nectar before the experiment continued to attract them. Thus, science once more and from another starting-point arrived at the conclusion we have long been familiar with from the work of Lunino bee-keepers.

The influence of competing honey-plants can be easily removed. The time for which a certain smell is remembered is limited, as is well known from previous experiments. To eliminate other influences, it is enough to make bees forget the location and scents of the competing plants.

With this aim in view, bees were kept in the hive for two or three days before being released for working clover. All this time they received fresh food in the shape of clover-scented syrup. In this experiment the number of bees visiting clover was 10 to 20 out of a hundred. This was the first more or less satisfactory result: every collective farm has, or may easily set up, an apiary of 10-12 hives and trained bees from such an apiary can successfully pollinate a medium-sized clover plot.

When all this was clear, the following experiment was carried out at a collective farm in Firsovo District, Kalinin Region.

Yellow Caucasian bees fed on clover syrup and black Central-Russian bees fed on heather syrup were released on July 30. Observers registered 2,025 yellow bees on red clover and 2,250 black bees on heather at the same time. When the bees returned to the hives at night they were shut up for retraining. Black bees received red-clover syrup and yellow bees—heather syrup. On August 3, the bees were released once more and the observers counted 2,875 yellow bees on heather and 2,837 black bees on clover.

This may be considered a tangible instance of bees obeying man. Then the time came for the bees to work clover fields in good earnest.

To send bees to clover by feeding them with syrup is not a difficult operation. Half a kilogramme of sugar is dissolved in half a litre of boiled water; the syrup is cooled to room temperature and fresh clover corollas with their perianths removed are put into it. In two hours the syrup acquires the scent of the flower and is poured into feeders, 100 grammes in each.

The drop of honey which the returned forager regurgitated and home-bees received into their proboscides, the drop of aromatic nectar which the forager brought in her honey-crop as if it were a corked scent-bottle, has become 100 grammes of aromatic syrup in the feeder.

Hundreds and thousands of bees suck this man-prepared nectar and receive from it an impulse and direction to flight. The feeders are put into the hive early every morning during the blossoming period of the plant which the bees are sent to work.

Upon the elaboration of this method it was no longer necessary to bring cumbrous apiaries from 60 to 120 hives each into red-clover fields: a dozen trained colonies did the work very successfully. One colony trained on clover might be said to do the work of ten.

Some time later methods were found of eliminating old bees from the colonies. Old foragers visiting other plants recruit young bees about to take field duties. That is why colonies made up exclusively of young bees which had nothing to forget were more teachable and more easily trained by means of scented syrup.

Meanwhile research in this field was continued.

It was astonishing how many questions remained unanswered when the whole matter seemed to have been settled and the method had met with recognition.

The feeder of syrup sends to the field of a given culture 10 to 20 per cent of the forage force, instead of 1 to 2 per cent that fly there without this impulse. That is good but it is not enough. Indeed, only two bees out of ten obey the will of the bee-keeper, the rest escaping the control of man. But bees themselves know how to utilize all their forces—we have mentioned the ability of the colony to mobilize all its foragers, including bees not yet ripe for field duties when a heavy honey-flow is on. Why cannot man train bees to do this at his bidding? Soviet apiarists are persistently working in this direction.

Now, to sum up.

It was commonly believed at the beginning of the present century that bees cannot pollinate red clover. Darwin adhered to this opinion and so did all the world's leading botanists and entomologists.

By proving that all strains of the honey-bee visit blooming red clover Soviet scientists have turned a new page in the cultivation of this important agricultural crop

Today, this flower "locked" to science and actually inaccessible to short-tongued bees has been "unlocked" by Soviet science, which makes bees visit red clover despite the fact that in pollinating its flowers bees do not obtain a load of nectar.

As far back as 1939, foremost bee-keepers reported at the All-Union Agricultural Exhibition in Moscow that, with the help of trained bees, they had obtained a twofold and threefold increase in crops of precious clover seed.

Training bees became very popular. Fed with scented syrups, bees visited vetch, coriander and even potatoes, which, left to themselves, they never visit.

Collective-farm apiarists in the Crimea were delighted at the sight of trained bees carrying loads of vine pollen. Vine-growers could scarcely believe their eyes: fed with *Chaush* syrup bees found this variety out of dozens of others and worked it.

Plant-breeders, botanists, agronomists, and apiarists from all over the Soviet Union were interested in this work, young nature-lovers working at school apiaries asked for advice and instructions. The news of the discovery of Soviet science told apiarists about the achievements of bee-scientists and practical bee-keepers in controlling the field activities of bees.

It is interesting to recall some data bearing on the problem. The discovery that insects pollinate flower plants was made in the sixties of the XVIII century. The "dance" of foragers was first described in 1823. A hundred years later—in 1923—this phenomenon was tentatively explained as the dance of scouts calling other bees to collect nectar. And in 1930 for the first time in history a group of bees flew in the direction indicated by the agronomist from the Experimental Station in Moscow.

By directing the flight of their bees, Soviet scientists have written a new page in the history of controlling living nature on our planet.

A Double-Locked Flower. A Wild Bee and a Bumble-Bee on the Keel of an Alfalfa Flower. The Paradoxical Effect of Training Bees for Pollinating Alfalfa. How Bees Can Be Made to Collect Pollen. Cloning and the Interdependence in the Selective Breeding of Bees and Alfalfa.

Alfalfa seeds are needed by Soviet agriculture in as large quantities as clover seeds. Like clover, alfalfa sown in a mixture of grasses is an excellent agent for restoring soil structure. Like clover, alfalfa can yield in many districts large crops of excellent green fodder and hay for the animals. As specialists in zoology say, this grass is of high nutritive value. Soils on which clover has grown produce exceedingly high crops of flax, and soils left over from alfalfa are very good for cotton. Clover saves soils from exhaustion by flax, and alfalfa lowers the level of subsoil waters, thereby preventing the salinity of irrigated lands, the veritable scourge of cotton-growing.

But we have mentioned above that it is no easier to obtain seeds from alfalfa than from clover.

At first glance alfalfa seems an extremely capricious plant. Sown pure, without admixture of other grasses, alfalfa in many areas produces no seed to speak of, even when the individual plants are well-developed and do not crowd one another.

This peculiarity of the behaviour of alfalfa, inexplicable at first glance, is explained by the fact that its seeds are spread neither by the wind nor by animals. They fall on the earth around the mother-plant.

But what is the connection between these phenomena? There *is* a connection, as will be seen presently. Since in a field of thickly-sown alfalfa each plant is surrounded by plants of the same perennial alfalfa, seeds are of little use for the species. It is more advantageous for the survival of the species to have the substances that might go into the production of seeds accumulate in the root collar as stores of food for the shoots of next year.

The correctness of this explanation is corroborated by the fact that in rarefied patches that same alfalfa sets seed even if surrounded by other grasses and crops. It prepares seeds for spreading over areas occupied by other species.

There is another, equally queer, aspect of alfalfa's behaviour. It is justly considered an exceedingly moisture-loving plant. But seed set is obtained in this moisture-loving plant only if soil humidity begins to decrease from the time of the appearance of flower-buds and continues to decrease until the beans are ripe. No sooner is alfalfa watered during this period than the plant ceases to supply with food those old shoots which have flower-buds, flowers or seeds and starts sprouting anew. This, too, is detrimental to high seed crops.

The honey-bee often takes nectar from an alfalfa flower in an "illegal" way, from the side. The flower remains untripped and no pollination occurs

The worst thing about alfalfa is, however, that it may be sown correctly and watered in time, and still it will fail to produce seeds, because its flowers often remain unpollinated even in the best of weather.

The construction of the flowers of alfalfa—blue, yellow and multicoloured (hybrid) — is exceedingly complex and fine. An opened bud of any other plant is ready for pollination and either the wind or some insects attracted by nectar will bring to the stigma pollen from other flowers. A ripe flower of alfalfa seems to be locked and it is not easy to describe the construction of the lock.

The so-called style consisting of nine accrete stamens and one free stamen, forming together a tube pierced by the pistil, lies at the base of the biggest and softest petal (the vane), covered all over with a thin and almost transparent tissue of other petals (the keel).

The style is highly resilient; it seems to wish to break itself away from the keel in which it is kept fast by the two ribbon-like appendixes of the wings, each fixed in small grooves on both sides of the keel.

The force with which the style tends to break away from the keel has been measured and found to be equal to a weight of five grammes.

This structure made up of tender petals and holding the live spring of the style as in a cocked gun is simply amazing! But what is its function? The stamens carry anthers. The anthers burst in the bud and cover the stigma with pollen. We may think this is an instance of self-pollination, which is so harmful to plants, but we would be wrong. No selfing can take place here, because the surface of the stigma is covered with a mucous membrane through which not a single grain of its own pollen is able to sprout. To effect fertilization, the membrane, or part of it, must be removed from the stigma. After this, pollen (preferably foreign pollen) can sprout into the style, reach the ovary at its base, and set seed.

Thus we see that the ripe flower of alfalfa is actually double-locked and if fertilization is to take place, the style with the pistil must be freed from the keel and foreign pollen must be placed on the stigma freed of its membrane.

A bee flies to an alfalfa flower on a hot, sunny day. This is usually a wild solitary bee—a Halictus, a Megachile, an Andrena, a Melitta or a Melitturga.

The bee has evidently come from another flower: it is covered all over with pollen sticking to its hairs and making it like a baker covered with flour. Settling on the keel, the bee thrusts its proboscis down the corolla, pressing its head against the vane and its hind legs against one of the wings.

After this, things develop so rapidly that a description of them has become possible only in the result of long and persistent observation.

The appendix fixed in the keel is moved aside and freed by the insect's hind legs while the resilient style escapes, and the whole system of "locking" the flower is destroyed.

The style glides away from its resting place, covers with pollen the bee which deftly dodges the blow, and springs upwards at a tangent striking against the vane.

The flower is opened, or tripped. The mucous membrane of the stigma is left on the vane; the bee dives deeper towards the base of the style for more nectar and leaves grains of foreign pollen on the stigma.

All these events take place within a second or two, but it took years of persistent work to establish the co-ordination and function of each detail.

Sometimes instead of a wild bee a bumble-bee alights on the keel. This big ponderous insect acts in like manner. But it makes its way to the nectar with such greedy haste that after its visit the flower is completely ruined: the keel is often torn off its base, the style displaced and the vane broken. Nevertheless, the seeds set in spite of these ravages.

Thus, if areas under alfalfa abound in wild bees and bumble-bees, other conditions being favourable, a crop of seeds can be relied upon.

In a hectare of well-sown alfalfa about 50,000,000 flowers open every day for about a month, and each can be fertilized completely after at least two visits of bees. Consequently, thousands of pollinating insects are needed to fertilize all the flowers on a hectare of blooming alfalfa.

Progress in agriculture means extension of area sown to various crops, including alfalfa, which, in its turn, leaves less and less uncultivated land where wild insects build their homes. As a result, the lack of wild insects is already noticeable in localities of intensive farming and is becoming an obstacle to high seed crops in the most advanced economies.

This is one of the contradictions arising from the clash between nature and production. But the latter has a satisfactory method of training bees at its disposal and there would seem to be no reason why bees should not be sent to work alfalfa as they do red clover.

And so hundreds of experimenters at collective and state farms, scientists at institutes and agricultural stations started feeding bees with alfalfa syrup.

Bees trained by this method began visiting alfalfa fields and working one flower after another. The data concerning the number of trained honey-bees on alfalfa raised sanguine hopes in the hearts of the experimenters, but when autumn came seed crops were pitiful even on plots known to have been visited by trained bees.

The results of threshing seed alfalfa were contradictory and many specialists came to the conclusion that the honey-bee cannot be of any use in this respect while some went so far as to state, on the basis of their experiments, that training was even harmful. To bear out their statements, they quoted data about crops obtained near apiaries where the number of bees was obviously greater than on plots farther from apiaries and consequently less visited by honey-bees. However, in the former instance the seed harvest was smaller.

But it is an instructive fact that in the given instance smaller crops were proof of the effectiveness of training.

This paradoxical conclusion is explained by the behaviour of foragers on alfalfa flowers.

Honey-bees, which are somewhat bigger than wild bees and much smaller than bumble-bees, usually alight not on the entrance to the flower but on the side—on the vane or the pedicle—without touching the appendixes of the wings. When bees settle on the keel they press against both wings simultaneously. Both in the former and in the latter instances they reach the nectar at the base of the vane with their proboscides in such a way as not to set in motion the mechanism opening the flower. The flower remains "locked"; but its wealth is taken from it in spite of the lock! Flowers deprived of their nectar offer no attraction to insects, none come to them, and they remain unfertilized.

Thus, instead of raising seed crops trained bees lower them.

If, however, a clumsy bee happened to unlock the flower by touching the wing appendixes with her legs, the style jumped up with the force of five grammes, gave the insect a blow, sometimes jamming her tongue or leg. The bee would extricate herself out of the trap with difficulty and then rest for a long time, perhaps rubbing the injured part of her body. And there are observations proving that a bee, once trapped in an alfalfa flower ceases to visit this perfidious plant.

At the same time some bees "trip" alfalfa flowers in the most orthodox way.

It was long before scientists succeeded in unravelling this tangle of contradictory evidence. This was made more difficult by instances of spontaneous opening of alfalfa flowers, mostly dark-coloured ones. As a rule, this is observed on hot days and is supposed to happen owing to difference of temperature in the different parts of flowers. This is very likely, since a pencil of sun-rays directed through the magnifying glass on the lower part of the keel opens the flower immediately.

Flowers may open on windy days owing the mechanical action of the wind, as a result of swaying, shaking or knocking against other plants.

Tripping through mechanical means can be effected artificially by swinging a rope over the heads of blossoming alfalfa plants. Although a certain percentage can be tripped by this means, the work of trained bees on such plots does not result in raising the seed crop sufficiently.

This, too, is explained by lack of harmony between the anatomy of the flower and the body of the honey-bee: training induces bees to collect nectar which they can often reach (even in a tripped flower) without touching the stigma. Again the flower is unpollinated.

All this was very disheartening and a specialist in alfalfa defended a thesis proving that honey-bees are ineffective pollinators of this plant.

But scientists working at the South-East Regions Institute of Grain Economy in Saratov would not be reconciled to this hopeless conclusion.

"No pollination is effected because bees collect only nectar," the Saratov specialists said. "When, however, bees gather pollen from alfalfa they pollinate the flowers all right. Suppose we train bees to gather alfalfa pollen. True, we do not yet know any efficient methods of training for pollen; we shall try coercion."

Nine colonies of equal strength were taken into a large alfalfa field on the Voroshilov Collective Farm. All stores of beebread were removed from three hives; from another three hives beebread was also removed and feeders containing alfalfa syrup were put into them every day; three more colonies were left to themselves as control bees. In the entrances of the six hives under experiment were inserted pollen traps which allowed the passage of bees into the hive but cut pollen loads off their legs. In this way no matter how much pollen the bees gathered there was no beebread in the hives and the foragers had to fly afield for more pollen all the time.

That was a cruel experiment and it was clear from the start that the colonies experimented upon would suffer severely. The analysis of the pollen loads brought in by the colonies revealed the following state of things. In the first three colonies which had no beebread and did not receive alfalfa syrup a third of the foragers came with alfalfa pollen loads, thrice the number registered in the control hives. In the hives of trained bees, half the foragers brought alfalfa pollen, four times the number observed in the control hives.

The problem could be regarded as solved: the bees obeyed the agronomist's will and, in spite of all that had been said before, visited alfalfa pollinating its flowers and effecting a set of seed.

But a bee-colony deprived of beebread is doomed to a certain, if slow, death, and no one can expect its members, worn-out and scratched all over by the sharp edges of the pollen traps, to be able to collect a sufficient store of honey.

So it was quite natural that plant-breeders took up the problem of alfalfa pollination by bees. They had always known that in any alfalfa field there were plants somewhat differing in size, in minute details of the construction of flowers, in the force of the style, in the endurance and fitness of the locking apparatus in the keel and the wings. Bees were made to help scientists in selecting plants that were easy to pollinate.

In a hot-house one winter, several dozen alfalfa plants were cloned and each clone was grown separately. The idea was that plants within a clone differ but very little and may, indeed, be regarded as one plant. A bee may chance to visit one plant among many others but it would be very difficult to see her do it while bees visiting different clones might be more easily observed.

The clones were planted in a field and at their blooming period observations were held from 11 a.m. to 3 p.m. and each half hour the number of bees on each plant of each clone was counted, special attention being paid to flowers tripped by bees. It became clear from the outset that different clones attracted bees differently. On some clones only two bees were registered while on others over 60 were observed. The percentage of effectively-tripped flowers in the best clones was 86. The differences were observed on the following days as well and it became clear that some clones are worked by honey-bees as successfully as by wild bees.

On the other hand, observation of marked bees showed the number of foragers on alfalfa from different colonies. It was seen that colonies differed in the percentage of field-bees working alfalfa, in their behaviour on alfalfa flowers and in effectiveness of tripping and pollinating.

The possibility of co-ordinated selective breeding of bees and growing of alfalfa became a reality: selectionists would create new "bee-pollinated," varieties of alfalfa out of plants best suited for pollination by bees and "alfalfa-pollinating" strains of bees, out of colonies most effective in tripping alfalfa. Selective breeding throughout generations, reinforced by training on alfalfa syrup and pollen, must result in producing the desirable variety.

The combined efforts of plant-growers and bee-keepers, their daring searchings and jeweller like precision work will ensure stable and high yields of alfalfa seeds, thus solving a vital agronomical problem of today standing in the way of further progress in scientific farming.

WINGED ASSISTANTS

A Find in Der El Medin. Pollen Loads Collected by Bees Three Thousand Years Ago. More about Clover and Alfalfa Pastures for Bees. How Man Can Improve the Flowers of Fruit-Trees and of Sunflower. New Tasks of Selection in Plant-Growing and Bee-Keeping. Pollination by Bees as a Part of Selection and Seed-Growing Technique.

It has long been known from ancient papyri that when the Egyptians buried their dead they put dishes of various food, including comb honey in the tomb.

Documentary evidence is abundant enough but in none of the Egyptian tombs discovered during the last hundred years has a trace of either honey or comb been found. Special searches were unsuccessful and some Egyptologists began to doubt whether the hieroglyphs had been correctly interpreted. At last, in a tomb unearthed during the excavations in the late forties in Der el Medin, near Teben, Upper Egypt, which specialists consider as belonging to the epoch of the Nineteenth Dynasty (1350 B.C.) there was discovered a flat clay vessel—a kind of bowl—with the remains of a honeycomb.

Outstanding specialists were engaged in studying the traces of the piece of comb which had been in the tomb for more than three thousand years. It would take too long to describe all the devices the scientists employed to overcome the tremendous technical difficulties in analysing the blackened and caked honey dust. We shall confine ourselves to merely stating the most important results of the analysis. Grains of pollen were separated from the remains of honey, and their analysis enabled the scientists to establish the flowers from which the honey had been collected. At the same time the botanical composition of pollen which bees gather today in that locality was also established.

The ancient Egyptian honey contained almost no pollen of the plant varieties constituting the honey sources today, mainly field plants. The three-thousand-year-old honey was collected from wild forest varieties of trees and cultivated garden varieties.

The find in Der el Medin has once more corroborated the hypothesis that in former days the honey-bee used to be a forest insect, and that ages of domestication have made the insect feeding on nectar and pollen from forest varieties into one feeding on nectar and pollen from field varieties.

If we accept this hypothesis we shall be obliged to admit that bees, too, must adapt themselves to new, unaccustomed food. Then we shall better understand the numerous facts now coming to light of lack of co-ordination, lack of mutual correspondence between the nature of the honey-bee and the nature of many agricultural crops, much younger from the evolutionary aspect than the bee.

In natural conditions, both red clover and alfalfa grow in bigger or smaller groups scattered at some distance from one another. Usually the soil around them affords shelter to the long-tongued bumble-bee varieties, so ideally adapted to pollinating clover, while in burrows, in dry grass-stalks and under sun-warmed stones live various species of wasps and solitary bees, well adapted to the tripping and pollinating of alfalfa flowers.

On discovering the high nutritive qualities of clover and alfalfa, man started sowing them in the field, but, as has been justly remarked, artificial selection has little influenced fodder grasses whose development is proceeding in accordance with the laws of natural selection.

But by removing fodder grasses from their natural surroundings and sowing them on ploughed fields, man has unwittingly severed their natural connections. Red clover and alfalfa transplanted to cultivated fields have burst into blossom but the solitary bees and bumble-bees which used to pollinate them have remained in unploughed lands, unmowed glades and elevated spots in flood meadows. Then we must not forget that, under natural conditions, clover and alfalfa do not cover large areas, which is the case in cultivated fields. Besides cultivation provides for the plants going through the various phases of development more harmoniously than they can under natural conditions. That is why a cultivated field of flowering grasses requires an incomparably greater number of pollinating insects than the natural patches of these varieties.

The honey-bee is more and more becoming the chief pollinator of agricultural crops and has just begun to adapt itself to red clover and alfalfa as sources of nectar and pollen.

That is why the problems of raising short-tubular red clover and long-tongued honey-bees, of bee-pollinated alfalfa and alfalfa-feeding bees have become so urgent. These tasks of selection are dictated by the necessity of strengthening the weak connections between the red-clover species and the honey-bee species, between the honey-bee species and the alfalfa species.

The problems of bees on clover and bees on alfalfa have long been acknowledged as ones requiring speedy solution. And there rise problems of a remoter future: bees collecting pollen from maize and bees sucking up the sweet sap off a young ear of branching wheat, which has just emerged from its wrapping of leaves. Bees have been observed to do this occasionally and such observations must not be disregarded.

In summing up his conclusions on the variation of animals and plants under domestication, Darwin specially studied the problem of the so-called co-ordinating force, common in larger or smaller degree to all living organisms. He remarked: "When he (man) has produced any modification in an important part, he has generally done so unintentionally, in correlation with some other conspicuous part."

This remark has a direct bearing on the problem under consideration and is worth discussing.

We know that man cultivates some plants (cereals, leguminous plants) for the sake of their seeds, others (fruit-trees, berry-bushes, melons and water-melons) for their juicy fruit, still others (cabbage, tobacco) for their green leaves, others again (fibrous plants) for the fibre they yield, and others (beet-roots, potatoes) for their roots or tubers growing in the earth. Through selective breeding man has induced in the plants the development of those parts for the sake of which they are cultivated. The plant-breeder concentrated his attention on seeds or fruit, on tubers or leaves, he made them bigger, increased their protein, sugar, starch, or fat content, made them more juicy or more spicy, at the same time trying to make them lighter, transportable and easier to cook. In accordance with the aim the plant-breeder set himself, he increased the length of fluff on cotton-seeds, changed the composition of the milky juice of poppy, the make-up of tung and castor oil, the nicotine content of ripe tobacco leaves and the theine content of young tea leaves, the structure of flax and hemp stalks. . . .

Varieties born of nature and shaped by the conditions of life change for their own good, only to survive and to be able to flourish. With varieties cultivated by man in conditions supplied by him it is just the opposite.

Selection always has some aim in view, and when he carries out selection, man acts in accordance with that aim.

As Darwin said, speaking of live animals and plants, man "cannot observe internal modifications in the more important organs; nor does he regard them as long as they are compatible with health and life." He went on to ask: "What does the breeder care about any slight change in the molar teeth of his pigs, or for an additional molar tooth in the dog; or for any change in the intestinal canal or other internal organ? The breeder cares for the flesh of his cattle being well marbled with fat, and for an accumulation of fat within the abdomen of his sheep, and this he has effected."

Take for example fruit-growers who create new varieties of fruit-trees. What did they care about the fact that in some of their new varieties of, say, apples, the stamens were formed so as to prevent bees from reaching the nectaries or that pollen produced in the anthers was of low nutritive value to bees? The fruit-grower was interested in obtaining beautiful tasty fruit and frost-hardy highly-productive trees, and he has attained that aim.

There is a novelty of British fruit-growing, a variety called Bramley's seed hybrid, and said to be very good. Perhaps it *is* good, but observations have established that four times fewer bees visit the blossoms of this variety than the Pearmain Worcester that may happen to grow nearby. Orchardists complain that old foragers seem to warn young bees that they will find nothing attractive on the Bramley apple.

Actually bees are frightened away from the blossoms of this variety by the unusually long stamens forming a thick brush through which they cannot reach the nectary, as they do in other flowers.

The stamens in the blossoms of a new British variety of orange pepping also form a barrier through which a bee cannot penetrate inside.

Besides, several new apple varieties (including the Bramley) produce inferior pollen which bees avoid collecting. ·

The conclusion is that plant-breeders who do not care about the structure of the flowers of their fruit-tree varieties and do not discard varieties unsatisfactory in this respect make a serious mistake. ·

Like bees, fruit-tree varieties originated in the forest and their blossoms are, as a rule, better adapted to pollination by bees than those of many field plants. But one-sided artificial selection may cause, together with other changes, a change in the blossoms which essentially predetermines a relative low quality of the fruit and the relative short-livedness of the new variety.

The practice of plant-growing knows many such instances.

Sunflower, an important oil crop in the Soviet Union, is a cross-pollinator. Field cultivation of this plant was started but recently. And although selection in growing oil sunflower has been carried on for less than 100 years, there are sunflower varieties with a seed oil content reaching almost 50 per cent, so that each seed is nothing else but a drop of oil in a shell.

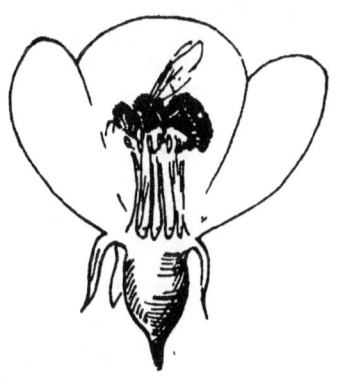

Did the plant-breeder who developed this oil variety care about the flowers of his creation secreting little nectar and its pollen being of inferior quality? Was he concerned with the fact that the corolla secretes sticky gum fatal for insect pollinators? He was interested in nothing but the high oil-content of the seeds, the sufficient size of the corolla

The blossom of the Pearmain Worcester (above) is well-suited to visits by insect-pollinators. *Below*: A blossom of a British apple Bramley's Seedling. Its compressed petals and the thick brush of tall stamens prevent bees from reaching the nectary

and the easy separation of the shell from the kernel. He has reached his goal and deserves praise for it!

But we know that the deterioration of conditions for cross-breeding influences a new variety in a bad way. Michurin's teaching of the vitality of plant organisms has enabled us to see in the flower of a plant an organ of reproduction, an organ for intensifying the vitality of a species. Consequently, the deteriorating conditions of cross-pollination will inevitably tell on the vitality of the progeny.

A bee's load of pollen is usually made up of pollens from numerous flowers of many plants, usually belonging to one particular species or even variety

We know from the chapter entitled "A Living Brush" of what importance for the fertility of plants are the number of visits by bees and the irritation of stigmas with the chitinous tests of bees. We know also how important for the viability, vigour and vitality of the progeny is the mixture of pollen brought from the stamens of numerous flowers visited previously.

The relative short-livedness of the best among self-pollinated varieties is a known fact. We have every reason to suppose that the progeny of cross-pollinators with deteriorated conditions for crossing will be shorter-lived and less vigorous than they can and must be.

It was quite natural that in experiments conducted by numerous research workers sunflower crops were invariably richer if additional artificial pollination with pollen mixture was resorted to. Seeds obtained through pollination with a mixture of pollen produced more vigorous and more viable plants yielding heavier crops.

What does all this mean? It means that without additional pollination the blossoms of plants are usually not completely pollinated. There are many special reasons for this in the case of different plants. In the case of sunflower, plant-breeders creating new—and better—varieties must develop such as will not trap bees, but, on the contrary, will secrete an abundance of nectar, which will attract bees in quantities ensuring complete cross-pollination, and, consequently, stable crops of highly-productive seeds.

If the problem of utilizing bees as an additional means of raising yields and improving the fruitfulness of seeds is to be tackled seriously, agronomists, plant-breeders, seed growers and bee-keepers must together revise a great number of agrotechnical plans.

Selectionists growing buckwheat, clover, peas, vetch and even potatoes, train bees to pollinate these cultures during their blossoming period. The seeds then are sown and bring in extra crops. When the need arises, the best are chosen out of the seedlings. Bees can be used in the selective growing of cotton, sugar-beet, flax, of vegetables meant for seed, and of trees, also for seed. Observation of bees foraging on blossoming grain crops gives ground for thinking that the possibilities of employing bees on selection and seed-plots are not limited to what has been described.

As has been noted above, Darwin explained the stability of heredity in the bee by the fact that it "feeds itself and follows in most respects its natural habits of life." But we know that from a species feeding on forest varieties bees have become one feeding on field varieties, from which we can infer that although bees seem to procure their own food, in actual fact they depend

A half-open flower-bud may contain nectar sometimes even in greater quantities than an open flower where it dries up and becomes so thick that bees are unable to suck it

for it on man who has radically changed the botanical composition of nectar and pollen pastures.

We are witnesses of the acceleration in the process of changing and renewing plant species which serve as food for bees. Now that methods of training bees have been elaborated, man is able to control the process of cross-pollination on large fields and plantations. At the same time the bee becomes a fine tool in the hands of plant-breeders improving plant varieties and intensifying their vitality, which opens up prospects for the planned creation of plant varieties which will be visited willingly by bees without coercion through training. It is easy to foresee that the renewal of plant species affording pasture to bees will proceed in future at a quicker pace and, for the first time in history, will assume a directed and planned character.

MORE ABOUT NECTAR PASTURES

Intervals in the Spectra of Blossoming Plants. Nectar Conveyer. The Place of the Apiary in the Collective Farm. When Forest Belts Start Blossoming. At the Source of New Honey Rivers.

Obeying the will of the agronomists bees will fly, as we have seen, to orchards and fields where they will pollinate the flowers of fruit-trees and berry-bushes, of sunflower and flax, of cotton and buckwheat. Floret after floret, bees will insert their proboscides into the narrow tubes of clover and into the locked keels under the vanes of alfalfa. A little later, each corolla will dry up, become brown and droop from the burden of clover seed and each keel will become a spiral-like alfalfa bean. And in autumn, collective farmers will rejoice as they watch streams of clover and alfalfa seeds flowing from the threshing-machines.

These seeds are necessary for the implementation of crop-rotation according to plan, they constitute an important link in the process of refashioning nature.

In drawing up charts of their fields and plans of crop-rotation, in marking the limits of various plots, agronomists and collective farmers bear in mind that the rise of yields cannot be based on chance pollinators: chance is an enemy of scientifically-conducted economy. That is why the drawings of the reconstructed collective-farm scenery include apiaries in the shade of the forest belts, where gaily-coloured hives stand in trim rows surrounded by hazel and acacia, honeysuckle, and lindens.

Now we must pause to think what will happen when the bees living in these apiaries háve been turned by training from the best honey-plants in whose flowers nature has prepared abundant stores of nectar and pollen. What will happen at the height of summer during the blossoming-time of plants producing the greatest quantity of pollen and nectar, if the bees have to work red clover? The fact is that out of the 160 kilogrammes of honey that may be collected from a hectare of blossoming clover, bees are actually able to gather not more than 6 kilogrammes. On every hectare over 150 kilogrammes remain inaccessible to bees. Thus, not only can clover not provide enough nectar to feed the bees that pollinate it, but as often as not it cannot compensate for the expenditure of honey on pollinating flights.

The way out may be to divide the foraging forces, so that while part of the bees are pollinating clover and vainly attempting to get the nectar hidden at the bottom of the florets, another part collects honey. Another way out may be the training of bees not only for directed pollination but also for directed nectar-collecting.

Bees need no training to visit the blossoms of white and crimson clover, of melilot, buckwheat, colza, mustard, raspberry and heather. But although bees come to these flowers of their own accord, feeding syrups smelling of white clover, melilot, mustard, etc., has proved an effective means of intensifying work on large areas covered by these honey-plants.

For a number of years, the flight activities and honey-collecting of several colonies fed scented syrups were compared with the performance of colonies fed pure sugar syrup, which is known to stimulate bees to flight.

The result was that bees given an extra ration, so to say, of pure *scent*, since sugar was received by all, collected much more honey: from buckwheat the difference was about 20 per cent; from crimson clover, mustard and heather, about 25 per cent; from white clover, colza and melilot almost 50 per cent, and from some other wild and cutivated plants as much as 100 per cent, and more.

The difference between the increase obtained from feeding one kilogramme of scented syrup and the same quantity of pure sugar syrup was one kilogramme.

Training bees for red clover also gave an increase in honey intake; while this increase did not exceed 10 per cent the crop of clover seed increased 150 per cent.

This new method of increasing the honey crop has vast prospects before it.

Every year a bee-colony consumes nearly 90 kilogrammes of honey and about 30 kilogrammes of pollen (monthly consumption has also been calculated). The bees consume all this food to be able to live, to perform home duties and field work, to secrete wax, and to feed the brood.

A colony that has collected 90 kilogrammes of honey has nothing to spare. An apiary must have sufficient areas of honey-bearing plants if the bees are to gather the 90 kilogrammes of honey required for their normal existence and dozens of kilogrammes extra to constitute the crop of the apiarist.

Some of the hundreds of species providing bees with food bloom in spring, others in summer, others again in autumn. The majority of plants yield both nectar and pollen but there are such as produce either the one or the other.

All such data are included in charts, or spectra, as they are called, showing bee-keepers the blooming periods and the nectar contents of honey-plants. If intervals occur between the blooming periods of wild plants, cultivated plants and crops in crop-rotation, or when not enough nectar is secreted to keep all the foragers busy, special bee pastures are set up near the apiary, where plants are sown which will blossom during these "empty intervals." With this aim in view, plots of fallow land are sown with early-blossoming buckwheat or phacelia, plots from which early potatoes have been gathered are sown with buckwheat; in some areas a mixture of crimson clover and Locus corniculatus is cultivated on apiary plots while on others, sainfoin is grown.

When sown on different dates, buckwheat and phacelia blossom at different periods. Sowing honey-plants at different times has proved a reliable method of extending blossoming periods and the honey-flow. Well-applied fertilizers increase the number of blossoms per unit of area.

Specialists, however, are not satisfied with one tried method of extending bee pasture, and are persistently looking for new ones. Owing to their efforts it has become known that feeding plants with some chemicals through spraying increases the nectar content in the blossoms.

Selection holds still better prospects in this respect.

Through crossing sunflower with Jerusalem artichoke a hybrid has been obtained with tubers sweeter than sugar-beet and stalks with a high sugar content attracting ants, wasps and bees.

By simply choosing the best out of old honey-plants nectariferous flora has been greatly improved. Thus there has been found a variety of black locust blossoming from April to September. Selectionists have named it Acacia semperflorens, i.e., ever-blossoming. The acclimatization of species new to a given locality, such as tulip poplar and other highly nectariferous plants, opens up vast prospects.

Collecting honey from these species is, however, a thing of the future: as yet we have not done everything we might to take a maximum of nectar from old species.

It is well known what a rich pasture for bees mixed deciduous forests are. And it has been proved that bees, too, are a beneficial factor in the life of the forest: where there are enough bees, the botanical composition of grasses growing in forests is improved and the seed crops of many useful tree varieties increase. Under these conditions the process of self-rejuvenation ensuring long life and prosperity to a forest proceeds more successfully.

Therefore, today a better use must be made of such excellent sources of nectar as linden, the maple varieties (acer platanides and acer campestre), oak, birch, ash, pear, apple, hazelnut, red elder, privet, cherry, and apricot.

From hazel and willow, bees collect pollen as early as March; pollen and nectar are collected from currant, nectar from willow and cornel, in April. All these bushes start blooming four or five years after having been sown. May is the blossoming-time of many grasses and bushes of the young undergrowth.

In the forest, bees are welcome to maples, black locusts and lindens; from a hectare of these trees a ton of nectar can be collected. Birches and poplars are early sources of pollen; oak produces pollen in June; then there are Eleagnus angustifolia and fruit-trees.

As forest trees grow, their crowns forming a thick canopy, they exercise a beneficial influence on crops in near-by fields, which yield more nectar.

Recent investigations in the zone of fifteen-to-twenty-year-old forest belts in the Stalin Collective Farm, Salsk District, have shown that the taller and bigger the trees the more nectar is secreted by sunflower, sainfoin and phacelia. Conditions for flying afield improve and the rate of working the flowers noticeably increases.

Thus, bees are employed in the field of legumes to facilitate the implementation of a scientific system of farming. In its turn, this system makes for further successes in bee-keeping.

Agronomists have made the following calculations for Sas District, Ryazan Region: after the collective farms of the district have completed planting forest belts and the bushes have started blossoming, it will be possible to increase the number of bee-colonies two and a half times with a corresponding increase in crops of honey.

In ten more years the trees will have begun blooming and the district will afford pasture to 15,000 bee-colonies which will produce up to 600 tons of honey.

These estimations do not take into account the possibilities of increasing honey intake through the systematic training of bees. If this is included in the estimation, the possible annual honey crops may reach 800 to 1,000 tons.

These are unprecedented figures for Sas bee-keepers.

When this is brought about honey-extractors in the collective farms will work at full capacity and honey rivers will flow from the apiaries in barrels.

This makes specialists in agriculture think that time has come to perfect the honey-bee for the performance of tasks in store for her.

A PEEP INTO THE FUTURE

THE SENSE OF TIME IN BEES

A Discovery on the Feeding Table. Bee-Clock and Bee-Calendar. Beehives in a Salt Mine. Flowers Open to Schedule. Sweet Lessons in a Bee-School, or How Bee-Keepers Made Foragers Carry Treble Loads.

"Who knows?" Professor G. A. Kozhevnikov used to say, "perhaps bees of the future will as little resemble the half-domesticated creatures of today as the draught-horse does the Kirghiz horse or the Cochin-China fowl the wild Himalaya fowl."

We see that he had in mind only the external characteristics—the structure, size and weight of the body, but selection will probably change not only anatomic details (which are of great importance, as the tongue length) but, first and foremost, many characteristics of the behaviour of the bee. In this respect there are many things to perfect in the bee, among others, precision and accuracy.

To explain our point, we shall have to make a slight digression.

Several years ago bees were being trained at an experimental apiary. There, a feeder of sugar syrup was put near a hive under experiment every morning. The feeder was a flat vessel with a diminutive raft from which the bees could suck the syrup. Covered with a net, the feeder was carried to a remote corner of the garden and placed on a little table. The bees were then released and when they came back they were marked in various colours. From the first day of the experiment the marked bees flew busily from hive to table and back and the observers at the table and the landing-board noted their numbers in record books.

The data from the books were used to draw up schedules of the work performed by each forager, showing the number of visits and the time spent on flights. In this way the memory, the velocity of flight and diligence of the bees were studied.

Once the bee-keeper in charge of the experiment happened to arrive at the apiary somewhat later than usual. At the time when the feeder should have been set out and the bees should have been carrying their loads to the hive he was going along the garden path. He approached the feeding table which was 500 metres away from the apiary and to his great surprise saw bees on the table. Those were not chance visitors but his marked bees. They were crawling all over the table in search of the feeder, flying up and again settling on the table.

Why were there so many bees on the table on that particular day when the experimenter was late and at the particular hour when the experiment should have begun? Why are there no bees on the table at other hours of the day when no feeder is set out? Why were there so many bees now that the table was empty? What had brought them here? If this was not mere chance, then the bees must have remembered, so to speak, the hour when the feeder was out.

But could that be? Were bees capable of remembering not only the place but the time when the feeder was put on the table? A simple experiment was started to verify this surmise. Bees were trained to come to the table at eight sharp at which time the feeder was set out. At ten sharp it was taken away.

This was done for ten days.

Marked bees came to the syrup in a body. At eight o'clock on the eleventh day the feeder was put on the table as before, but it was empty. At first many bees came to it but some time later their number began to decrease. The most persistent returned to the feeder up to ten o'clock.

More experiments were staged to find out whether bees could be trained to come at different hours of the day, in the morning, early and late in the afternoon. The same experiments helped to ascertain whether bees distinguished and remembered different time periods—one hour, two and three hours.

Bees invariably manifested the ability to be exact, and the more concentrated the syrup was, the sooner they displayed their punctuality.

A group of bees was trained to take syrup from a feeder from ten o'clock to midday, and almost all the bees from this group would come to the feeder even when it was empty. Two hours after the beginning of the feeding-time the majority of the bees stopped coming even if the feeder was left on the table after the expiration of the time specified.

There was no doubt that the bees had a time-sense.

It was established that one and the same group of bees could be taught to come to the feeding place at fixed hours, two, three, or four times a day. Bees could observe two-hour intervals with ease.

Almost all animals, birds and fishes generally feel the feeding-time very well. Tamers of wild animals and zoo staffs as well as animal-breeders, lovers of birds, people having aquariums at home and engaged in fish-breeding can tell many an interesting tale about their charges' time-sense. But the definite manifestation of this sense in insects seemed astonishing.

But why do time-trained bees not participate in collecting nectar from other sources? Why do they ignore the circlings and tail-waggings of the dancers announcing the discovery of a rich source?

New observations of the behaviour of marked bees have given an answer to this question; time-trained bees spend their "free time" in the remotest corners of the hive, while the dances of the scouts are performed, as a rule, in the centre of the brood chamber or near the entrance. That is why bees trained to forage at a definite time do not see the dancers.

Later it was decided to see whether bees would come for breakfast at nine o'clock in one place—in the garden—and for dinner at five o'clock in a glade.

After seven days' training bees once more proved that they knew their time and place. The conclusion drawn from this experiment was checked several times with invariable success. True, some of the bees came for their dinner after visiting the breakfast place and finding it empty, but none missed a feeding-time.

Subsequently a series of experiments were held to compare the tenacity of bees' memory of time and place. Bees coming to a right place at wrong hours proved that they remembered place better than time.

Similarly, it was established how long the bee can remember feeding-time. On the thirteenth day after training was discontinued, no marked bee came to the feeding place at the right time. Then the influence of counter-training was studied and bees trained to feed at one time were retrained for another feeding-time. In these experiments bees accepted the second time-table on the third day. Not all the bees, however, behaved in the same way. In one and the same colony there were bees punctual to the minute and "absent-minded" bees that came either too early or too late, bees that confused the times or forgot the place. But there were so few of them as not to affect the general conclusion.

This general conclusion was that, as a rule, bees remember time. Experiments showed that the bees did not fail to fly to the feeder in due time even after taking some honey from the comb. So it was not hunger that prompted them. In this connection it was necessary to ascertain what it was that the bees remembered: the moment of feeding or the intervals between them.

When food was given not every twenty-four hours but at intervals of, say, nineteen hours, the time of feeding fell at different hours of the day. These experiments showed that bees did not "understand" or accept nineteen-hour-long intervals. So training for different hours of the day at equal intervals was a failure. Slight variations were introduced into this experiment: bees were fed regularly between eight o'clock and midday on even days, the interval between the commencement of the feeding-time being 48 hours.

The results were not quick to manifest themselves, but after three feedings, on the seventh day of the experiment when no feeding was due, bees were punctual in visiting

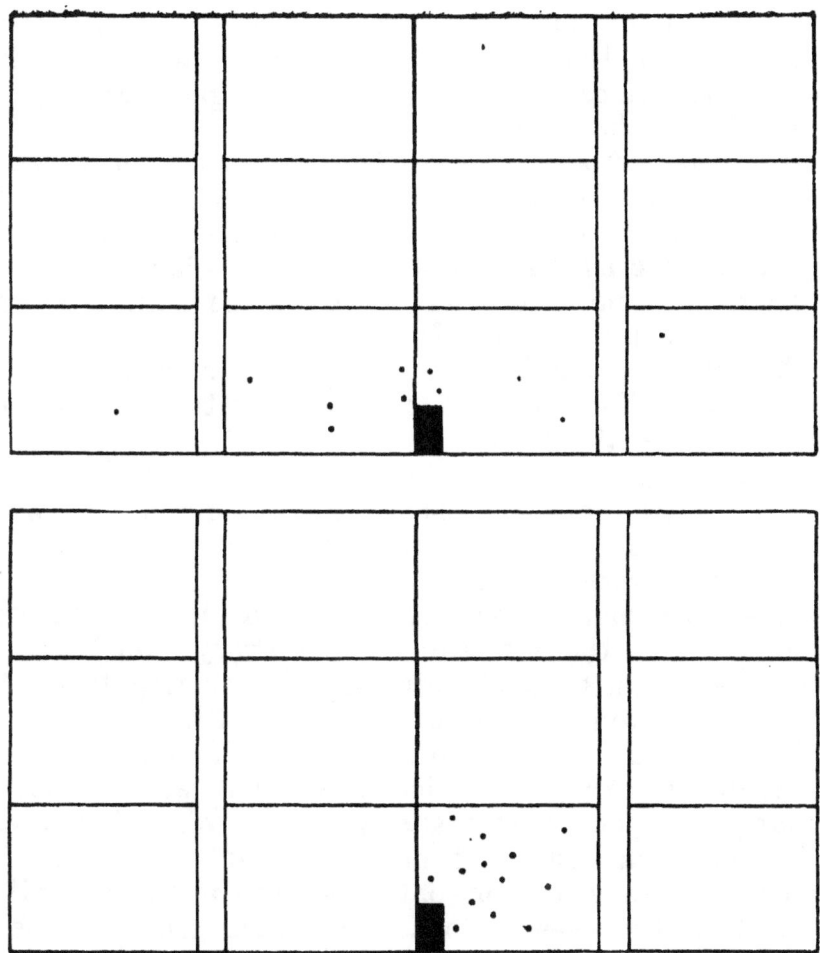

We see 12 comb surfaces of a six-frame flat observation hive inhabited by bees which were trained to take food off feeders between 10 a. m. and 1 p. m. The dots designate spots in which marked bees were discovered at 7.25 a. m. (above) and 9.15 (below). From this it is clear that as the time for departing drew nearer, the bees collected at the entrance

the feeder between eight and midday disregarding the fact that it was an odd day. Thus it became clear that bees live according to the natural division of the day into twenty-four hours.

This might give rise to the idea that bees felt the time by the sun, sensing either its height over the horizon or the slant of sun-rays. Accordingly, the experiments were held in a light-proof chamber with an electric light fixed in one place. There had to be a light, because bees do not fly in darkness.

In these new surroundings the bees behaved as usual. The experiments there were conducted both during the day and at night and the bees came to the feeder at the beginning of the feeding-time and stopped flying when it was over.

There was no direct influence of the sun, but what kind of influence was there?

The electro-conductivity of the atmosphere? Or some sort of rays? For both are connected with the sun and, consequently, with time.

Perhaps, after all, the idea that bees can, in some way or other, sense these invisible and inaudible signals which man can decipher and record only with special apparatus, is not so very absurd.

The experiments were again staged in a light-proof chamber where the air was ionized every two hours in order to extinguish the electric signals coming from the sun. But this did not confuse the bees either.

In the ionized atmosphere of the chamber, too, the bees regularly came to the feeder when it was out and as regularly stopped their visits when it was taken away.

So this idea had to be discarded. But before arriving at the final conclusion, one or two points had to be clarified.

What if the bees' "clock" was some kind of rays or currents unknown to man? To find this out, bees had to be removed from the surface of the earth where they might be influenced by these rays or currents.

An unusual load—beehives—was taken down a salt mine in a cage.

Electric light was switched on not to be extinguished before the end of the experiment in an empty mine a kilometre away from the workings. Here, at the depth of 180 metres a constant temperature of 16 to 17°C. was maintained. The entrances and ventilators were shut tight, to isolate the place of experiment from the outer world. There was enough air.

Above the hives were the underground vaults sparkling with salt. The sun could exercise no influence on the bees which were cut off from the world on the earth's surface and from its signals. Perhaps here the bees would lose their time-sense?

Training lasted for a fortnight. On the fifteenth day, the first day of the experiment, the observers at the feeding table saw the bees behave in the dead stillness of the forsaken salt pit just as they had behaved under the bright sun among living nature: at off hours no bees were seen at the feeder and during feeding hours the little raft on the bowl was covered with bees.

So one more hypothesis was discarded.

New hypotheses were studied and discarded one after another as, after seven to ten days of training the bees came to the feeder punctually at fixed hours.

All these experiments suggested the conclusion that the evasive "clock" should not be looked for outside the organism of the bees.

Combs with sealed brood were placed in a light-proof chamber with the necessary temperature and humidity. Bees born here did not see the sun or the sky, knew nothing of the change of day and night. These bees did not come in contact with old bees whose behaviour they could have imitated. And, like bees born in a crowded hive, these bees learned to come to the feeder at set hours. It was clear that there was no sense in continuing research along old lines.

All the experiments positively showed that the sense of time, just like flying or the number of joints in their feelers, was a natural characteristic of bees.

But then it had to be ascertained what benefit bees derive from their time-sense which seems to be a firmly-rooted instinct. What is its biological function?

Botanists and naturalists have long been aware of the fact that many plants in each locality open at a certain time. Linnaeus utilized this phenomenon to construct a "flower-dial" which enabled him to tell the time fairly accurately by day and by night. Interesting new data on this subject were collected at the Experimental Station in Tula.

On bright sunny days, observers came to experimental plots and with the help of graduated glass rulers measured the height to which the nectar rose in flowers at different times of the day. They also weighed pollen from the stamens. This was a very delicate and tiring work but it ena-

Pollen loads from garden poppy often are quite black

bled the research workers to make a discovery: the measurements showed that the quantity and quality of nectar secreted by the flowers of almost all plants are different at different hours of the day. At some hours there is much nectar and it is very sweet, at others there is less nectar and it is thin.

A schedule was drawn up for a great number of plants in which were noted down the hours when their flowers are richest in sweet nectar and ripe pollen, as well as those when the nectar is thin and the quantity of pollen small. This flower schedule was then compared with the data obtained by the study of bees' time-sense.

Biologists have long held that not only flowers are perfectly adapted (for their own benefit) to visits of definite insects but also that the insects themselves are equally well adapted to obtaining nectar or pollen from definite flowers.

Now this mutual adaptability and co-ordination, in addition to the data of anatomy pertaining to the details of body structure, have been confirmed by the mutual dependence between the blossoming time of plants and the time-sense of bees.

Here is an excerpt from the diary in which the results of experiments were put down.

"Ten blossoming poppy-plants were put under observation. The blossoms opened at 5.30 a.m. Out of the ten bees marked on poppy the preceding days, two came at 5.25, five minutes before the opening of the corollas; two came exactly at 5.30; three were somewhat late, coming between 5.30 and 5.32; two more were ten minutes late and one, a quarter of an hour late. One of the two bees that came before the time and three that were too late were young foragers on their second day of field work."

Incidentally, these observations, as well as the greater percentage of irregular cells in combs built exclusively by young bees, lead us to believe that young bees "learn" something from their older sisters. Perhaps this "instruction" is of the same nature as the effect produced by tapping one's finger-nails against the floor on incubator chicks which do not yet know how to peck. Whatever the case may be, the experience and working habits of older bees are a kind of "mentor" for the young bees and may prove an additional road towards controlling the development of the colony.

When the experiments were completed it was established that young bees were quick to learn and on the fourth or fifth day came with but a small time error.

This was observed on the flowers of 35 different species and varieties, including poppy, eglantine, rose, verbena, chicory, and others.

The observations showed further that flowers yielding nectar or pollen all day long are visited by bees from morning till night. The greatest number of bees visit such flowers early in the morning (when more nectar has accumulated during the night) and at the hottest hours of the day (when the air is dry, the water from the nectar has evaporated and its concentration is higher).

The shorter the time during which nectar and pollen in the flowers are available to bees, the more accurate the coincidence in time between the maximum food on bee pastures and the number of bees on them.

Thus it has become clear that time-sense enables the bee to reduce the number of idle journeys, to utilize to the utmost each minute of flight, to spend as little honey as possible on collecting nectar, to visit more flowers and, consequently, to increase the colony's food stores, thereby consolidating the basis of its growth and prosperity.

But, come to think of it, this very peculiarity of the bee prevents her from collecting more honey.

Within the flight range of each colony several species of honey-plants are usually in bloom. Therefore, each colony has several groups of foragers and each of these obeys the fixation instinct and works one species only. When the flowers of this particular plant cease to secrete nectar, the bees that used to visit it stop work for a time until "their" plant has resumed blooming or until they have forgotten it.

That is the reason why it often happens that at the time a new plant rich in nectar begins blooming, some of the foragers of a colony are busy collecting nectar from other plants, which are perhaps not so rich in nectar, but have started blooming earlier while others are enjoying legitimate "leisure."

With some bees it is a matter of several hours and with others—of several days, before they start working a new source. And loss of time means loss of honey.

In the chapter entitled "The Main Honey-Flow" we have shown how much a colony loses when its foragers have to change pastures.

Nearly the same may be observed when the foragers change times of flight, although in this latter case the picture is somewhat blurred. A change in the feeding time-table invariably resulted in a considerable reduction of the number of foragers coming at the right time and in greater deviations from it.

The experiments showed that a change in the place and time of feeding involved great losses for the colony.

Perhaps that is the reason of the shorter life span of "summer" bees, which we have treated in the chapter on prolongation of life. We see that this peculiarity of bee nature, too, has an adaptive character which in the given case manifests itself in an unexpected and unusual way. Indeed, the short life span of individuals and the rapid change of generations are determined and conditioned by a rapid change of place and time of foraging. This is how the change and expansion of the forage area take place in summer, rapidly changing with the succession of various blooming plant-species.

Thus it turns out that the naturally short life span of the summer bee is also, in a way, a necessary condition, a prerequisite for successful foraging and, consequently, a condition of the long life span of the colony as a whole. It is a case of "dying meaning living!"

A practical example of the results to which leads the foragers' natural fixation for the place and time of collecting food is furnished by the loss of linden honey in Lunino with which the reader is familiar.

One of these natural traits, namely, the attraction of the place for foragers, serves as the basis for the first and comparatively old method of training bees for the pollination of definite *parts* of a field.

A feeder containing scented syrup put into the hive gives the bees the impulse to fly. On leaving the hive they find feeders with that same syrup not far from home. When enough bees have settled on the feeders, they are covered with a net and brought to the spot of the field to be pollinated. The net is removed and the bees released. It remains only to pour more syrup into the feeders in the field which will attract more bees to the spot. When a sufficient number of bees are flying to the feeders, these can be removed.

A. N. Nevkrytov, a Ukrainian biologist, has cleverly perfected this method of training bees.

He used to put a feeder into the hive at night and take it away early in the morning, half empty, together with the bees in it. Covering the feeder with a net he carried it to the plot where pollinating work was to be performed, and then put the feeder down and removed the net.

For several days the feeders in the field were refilled and the bees were given syrup in the hive, which completed the foragers' "territory training."

The foragers were free to describe their "figures eight" on the combs, conveying through the rhythm of the gyrations and waggings the exact location of the feeders and the distance to them. On coming to the place, the recruited bees acquainted with the smell of the syrup would find it only in the flowers which they would start working accordingly.

In this way man has utilized bees' instinct making them constant to the place of foraging.

And this is not all. The bee-keeper N. A. Solodkova reasoned in this way: "If bees could be trained for two forage areas simultaneously, the foragers could change places quickly and this change would cost less."

Bees remember well enough two exact hours and places of feeding. Can they remember two smells simultaneously?

One fine day when bees came to the feeder with the accustomed anis syrup they were covered with a net, marked red and released. The next day the bees had a choice between a feeder with anis-smelling syrup and a camphor-smelling one. Of course the red bees took no notice of the feeder smelling of camphor and visited the anis feeder.

And then N. Solodkova took it away.

This caused great confusion and hubbub among the bees. It took the unwilling anis foragers a long time to reconcile themselves to the smell of camphor. Still some of them filled their honey-crops with this strange syrup; such bees received additional yellow marks.

Training to two syrups was continued for several days. Incidentally, this experiment put a finishing touch to the time-sense experiments.

While experimenters were busy establishing whether bees possessed time-sense, all the experiments proved convincingly and beyond doubt that bees had it and preserved it under all conditions whatever. The general impression was that no external conditions could influence bees' time-sense in any way. Indeed, in a mine, deep underground, bees foraging not on flowers, but on feeders, reared in the mine and without any old bees that might "teach" them things known on the surface, were true to their inborn punctuality.

After all, there was nothing surprising in the discovery made during the experiments we are speaking about. It was not the experimenters' aim to change, to reconstruct the nature of bees. All the experiments, from first to last, aimed at establishing how firmly time-sense is rooted in bees, and the work, accordingly, was carried along passive, speculative lines.

No sooner, however, had an experimenter set himself the active task of disturbing, eliminating this sense, and not merely of proving that bees had it, than this new task helped to vanquish nature and showed how bees could be made to desist from following a natural time schedule.

On arriving from the hive the bees found in the accustomed place now an anis feeder and again a camphor feeder, which were changed at irregular intervals. Finally, the bees obeyed the experimenter and, disregarding time, came to the feeding place and took either of the two syrups that happened to be set out. After the bees had adapted themselves to the new conditions, feeding in the old place was discontinued and at a little distance from it were placed anis and camphor feeders side by side.

The ensuing events were to show how bees remembered two smells. The picture became clear from the outset: 242 red-and-yellow and 11 red foragers on the anis feeder and 231 red-and-yellow and 4 yellow bees on the camphor feeder proved that the bees remembered the two smells equally well.

A year later N. Solodkova successfully trained bees for three smells. Her bees ceased staying in dark corners of the hive, for at any time of the day they were recruited by red, yellow, and white dancers and by dancers marked in two and even three colours.

This was a new and great achievement in controlling the four-winged.

THE END AND THE BEGINNING

Unfulfilled Promises and X-Rayed Cripples of Weismannists. A Ray of Light in the Dark Hive. Bees in Earthen Nests. Variety Sucked with the Proboscis. The Least Obedient of Nature's Beings Will Be Man's Creation.

What should contemporary followers of Weismann do on learning about bees' sense of time? The teaching they profess holds that any character, any hereditary trait is determined by a particle of the hereditary substance—a determinant, a gene. Departing from this premise, Weismannists should be interested, first, in what chromosomes the gene controlling the time-sense may be hidden;

whether there is one gene performing this function or several, and if so, how many and what names should be given them.

That is the method they usually followed.

They spent years drawing up purely speculative charts of bees' chromosome complexes and guessing where the gene of swarming might be hidden, what particular spot of the chromosome harboured the gene of anger, and what combination of genes contributed to greater productivity.

Proof of the might and potency of their theory they saw in advertisements in which golden bees, exceedingly beautiful and producing unheard-of quantities of honey were claimed to be bred according to the latest prescriptions of Morganist science. They themselves dreamed of being able to breed bees in thermostats after such prescriptions.

Firmly believing that each body is merely a repository of the hereditary substance, proof against all external influences, they put in cages select queens and drones proclaimed to be sole bearers of the best bee genes and living "hereditary substances." True votaries of their theory, the Weismannists flatly refused to admit that the sexless nurse-bees rearing the larvae could change the heredity of the bees.

To admit that would have been to proclaim their theory untenable.

Not a single promise of the Weismannists was fulfilled. They failed to breed bees with longer proboscides or exceptionally large honey-crops (such bees could gather more honey), or bees with wider-spreading wings (such bees could fly better), or again, bigger bees (such as could pollinate certain closed "irregular" flowers).

In their search for magic ways of perfecting the bee, some Weismannist treasure-seekers decided to use X-rays for the purpose.

They started treating queens with X-rays, although none knew what good this might do and how the idea had arisen. After the treatment the queens became either sterile or filled the combs with eggs from which only cripples hatched.

A hapless Weismannist spent several years in his laboratory counting the number of hooks on bees' right and left wings. He summed up his findings according to the approved rules of variation statistics and arrived at an unerring conclusion: the number of hooks on the right and left wings of bees is usually the same and cannot serve as a character for selective breeding.

Life and the achievements of progressive materialist biology have done away with all this rubbish and showed that only in accordance with the laws of living nature, acting in unison with nature, can the bee-keeper successfully direct the characters and properties of bees, through specially created conditions towards the set goal.

It was recently established that a female pigeon reared in perfect isolation will start laying eggs only after seeing at least her own reflection in a looking-glass. In absolute isolation a physiologically mature bird remains sterile, in spite of the fact that pigeons do not live in organized biological communities.

How powerful, then, must be the influence produced on each individual bee by the colony with its complex and all-embracing organization, with its multiple connections unifying in a living community tens of thousands of individuals.

The force of this influence can be seen from many organs of individual bees performing quite new functions in the colony. Has not the honey-sack become part of the communal stomach? Has not the ten-barbed sting become a weapon of defence of the whole colony? And what about the scent-gland which in the females of all other insects serves to attract the males, while in the honey-bee it strengthens the ties within the colony and contributes to a better organization of field work?

For each individual bee, including the queen and drones, the colony it was born in is its mother, nurse, mentor, defence and shelter. It is the colony that brings up each bee, determining her tongue length, the way she seals the cells, and countless other properties and characters.

Bee-keepers are used to talking of the "temper" of each colony. Colonies are known to develop rapidly from the first days of spring or to build to strength by summer, to be "gentle," "irritable," etc.

P. I. Prokopovich once told his pupils that "by their nature some bees are more diligent and others less." A diligent colony sends its foragers afield earlier in the spring and discontinues field work later in autumn than others. In the morning its bees are the first to leave the hive and come home later in the evening. The bees of such a colony fly noticeably farther than bees from other colonies, and work the flowers more quickly.

A highly productive colony must not only forage well, it must at the same time be economical.

Dozens and hundreds of little traits, peculiarities and properties make for economy in spending the collected store. Investigations of the time-sense of bees have established beyond any doubt that there are colonies which strictly adhere to the time of flights and colonies apt to confuse the foraging time. Training experiments showed that some bees were quick to follow the order given through food while others proved simply stubborn: no matter how well they were fed they were very slack in working the crop in need of pollination. During the experiments we have described above as a "geometry examination" it was noted that bees of some colonies easily distinguished between squares and triangles, while bees from other colonies confused different geometrical figures, some of them preferring the square, others the triangle. In these fine features of behaviour the differences were those of *whole colonies*. The characters determining the individuality of a colony are innumerable, and the more observant the bee-keeper is the more he sees. He distinguishes the colonies by the taste of the honey, by the behaviour of the bees during manipulation, by the quantity of propolis on the frames, by the way the bees build the combs. Some of the distinguishing characteristics are important for the bee-keeper, others seem to be of no importance. The bee-keeper sets apart colonies that produce much honey, are less inclined to swarming, winter well, and are disease-resistant, concentrates his attention on such colonies and tries to determine the conditions which shape the particular characters of the bees.

The study of bees in different latitudes is very important for bringing to light concrete conditions influencing the nature of the bee-colony. Since bees "have been transported into almost every quarter of the world" Darwin remarked that "climate ought to have produced whatever direct effect it is capable of producing." So Darwin thought that differences would inevitably become apparent upon comparing geographically removed strains of bees. He made this comparison and this is what he wrote:

"I procured a hive full of dead bees from Jamaica, where they have long been naturalized and, on carefully comparing them under the microscope with my own bees, I could detect not a trace of difference."

Now we know why Darwin failed to detect any difference: to see these differences, not dead bees, but living bee-colonies had to be compared.

In the severe tundra, summer is exceedingly short, but while it lasts the sun stays in the sky for many days running, shedding its light on vast expanses overgrown with willow-herb, heather, huckleberry, red huckleberry, and great bilberry. On Polar summer flight-days which are but few, bees work feverishly day and night without rest, carrying nectar into the hive and filling the combs with Arctic honey. The first autumn frosts begin in mid-August and from then the bees avail themselves of the warm hours, the number of which decreases day by day, to prepare the hive for the winter which lasts here for nine long months.

A distance of some 5,000 kilometres lies between the Polar apiaries and the apiaries in the Pyandza Valley, Tajikistan, where the winter rest of bees seldom lasts longer than one month and where February brings both the young leaves and the first blossoms.

And there are apiaries beyond the borders of our country, in the mountains of Colombia, South America, near the equator. Here is an apiary on a mountain slope at the height of 1,500 metres above sea level. The lowest temperature at night here is 14°C. and in day-time the highest is 28°. On cloudy days or during the rainy period the temperature varies from 16° to 22°. So all the year round the fluctuations of temperature are very slight. The mountains are covered with dense evergreen forests, palm-trees, and giant ferns. Some plant or other is always in bloom, and there are plants that bloom twice a year. Here, in the mountains, not only the temperature is constant; days and nights are always of an equal length, the day breaking abruptly at 6 a.m. and night falling as abruptly at 6 p.m.

These unusual conditions produce profound changes in bees transported here. They are shy of the bee-keeper and irritable, and in the second or third generation become so lazy that as soon as one or two combs are filled with the greenish fluorescent honey from lianas, orchids or coffee-bushes the foragers stop flying afield altogether.

To prevent the bees from degenerating in their habits, bee-keepers here order new queens every year from localities in which bees live with regular winter (or summer) breaks in the honey-flow. For this exigency fosters in bees the instinct for collecting and storing honey. We see that the saying we have had occasion to quote about greedy bees collecting honey and dying cannot be applied to the very same bees brought to the tropics.

The saying "every town has its own customs" can much better be quoted in this instance: the "avarice" and storing instinct of Northern bees, as well as the "laziness" and "carefree life" of Southern bees are brought about by the conditions of existence, they are a method by which bees adapt themselves to these conditions and a reflection in the laws of bee life of the climatic peculiarities of the regions bees live in.

As we have seen, this adaptation appears and becomes manifest in the progeny of transported bees at a very early date.

We have not yet discovered a direct explanation of the causes why Madagascar bees are black and why Asiatic albinos have white bands. We do not know why well-known Carniolan bees hardly need propolis while North-African bees cover their nests all over with a blood-red bee-gum. We do not know what has made the Caucasian bees so gentle (they go on with their work while the frames are being taken out of the hive) and the black Dutch bees, on the contrary, so fussy (as soon as the apiarist has removed the cover they start rushing about in a mass, like a flock of frightened sheep). We do not know why Italian bees are so good at fighting the wax moth that causes such a lot of harm to bees of other races.

But we have no doubt that every specific character has been developed under the influence, direct or indirect, of the environmental conditions.

Can there be any doubt that there is a direct connection between poor honey-flows and dispersed sources of honey on Cyprus and the industriousness of the dark-orange Cyprian bees, famous as fine honey-gatherers? Does not the capacity of the bees of Asia Minor to rear the brood all year round reflect the climatic conditions under which abundant loads of pollen can be collected in all seasons? Does not the inability of the Egyptian bees to store large quantities of honey reflect the slackening influence of the conditions under which the honey-flow lasts the whole year in the fertile irrigated Nile Valley? And, on the contrary, is not the influence of scarce sources in the oases of Sahara evident in the ability of the bees there to fly twice as far as our bees?

Careful observation of the changes in the behaviour of bees transplanted to new localities will certainly reveal many new facts important for understanding the correlation between the organism and its environment.

A close study of the numerous bee strains bred in the Soviet Union may furnish much new material in this respect. Some of these strains, such as the Far-Eastern, are resistant to foul-brood; others, such as the Siberian and the Ural bees, are frost-resistant; the Bashkir bees seal their cells with beautiful white cappings; the Kabakhtapin, Azerbaijan bees are very industrious; among the colonies of Armenian and Georgian bees there are individuals with proboscides eight millimetres long—a millimetre longer than those of the grey Caucasians.

An analytical comparison of the biology of bees living on different continents under different natural conditions is expected to throw light on some changes in bee nature wrought by different management within constant environmental conditions. The possibilities of such changes, too, are infinitely varied.

For a long time F. K. Babayev, a Ukrainian bee-keeper, has advocated the introduction into hives of comb foundations with larger cells. He himself made milling-rolls and produced his foundation on which bees built cells of six millimetres, i.e., 10 per cent larger than usual.

After using combs with larger cells for ten years Babayev obtained bigger and stronger bees.

The bee-keeper A. I. Igoshin has housed his bees in transparent, light hives with heated walls (bees usually cover cool glass walls with light-tight propolis). Here are some of the changes Igoshin has observed as a result of the light penetrating into the darkness of the hive. The cycle of development from grub to bee has become shorter by one day, life in the light has made the bee "earlier ripening"; the new heat regime in the hive warmed directly by the rays of the sun has protracted the foraging season. The bees in the lighted hive seem to have become gentler.

F. A. Ovchinnikov, a Siberian bee-keeper, has raised the entrance, making it midway between the bottom and top. This simple innovation has changed the temperature within. The bees winter more successfully and quickly build to strength in spring.

There are apiaries in some of the southern collective farms where brick and stone hives dug in the ground have long been in use. The everyday fluctuations of temperature in such hives are less marked than in ordinary hives and bee-keepers think that owing to this, other conditions being equal, the bees in "underground hives" produce more honey.

Larger cells, bigger hives, sun-lit hives and underground hives, all have shortcomings of their own which hinder their introduction on a large scale. But they have their merits as stages along new roads, as proof of practical bee-keepers' search for new methods of improving bee-colonies through changed management and environment.

Many of the seekers are blazing new, untrodden roads, forging ahead towards the goal by most unexpected by-paths.

Thousands of highly skilled bee-keepers, loving their craft with all their heart, bring their grains of truth to the store of knowledge of bee biology and discover new means of controlling bees.

But we have dwelt at length on the filters the colony has set around the nurse-bees and especially the queen, on the means by which it protects them from the influence of unstable external conditions. These filters are undoubtedly active under the conditions of changed management and environment, too. Consequently, these changes may fail to reach the reproductive centres of the colony through the filters, they may fail to influence heredity.

What, then, must be done to ensure their influence on the development of the coming generations?

Academician M. F. Ivanov, an outstanding Russian cattle-breeding expert, lays great stress on the decisive influence of food on shaping the heredity of animals. He is fond of saying: "Variety enters by the mouth."

The work with mixtures of honey and pollen carried on at the Institute of Apiculture has shown that if variety is introduced into bees through the proboscis the natural physiological barriers protecting heredity can be overcome.

We know that a bee-colony consumes on the average 30 kilogrammes of beebread and 90 kilogrammes of honey, i.e., about 3 times as much honey as beebread, a year. In the mixture of honey and beebread the ratio is one to one. The mixture is put into the hive and bees obeying the instinct forbidding them to leave honey outside the combs start carrying the mixture into the cells. Their honey-crops are filled with this unusually heavy food. As has been noted above, the bee organism is so constructed that in the honey-crop, which is part of the stomach of the community, food becomes purified, losing pollen-grains, since only pure honey must be stored for communal use.

If bees are fed with syrup containing an admixture of dyed pollen-grains one can easily see that the four-lipped stomach-mouth detains a large amount of pollen from the food in the honey-crop and conveys it in unusual quantities into the stomach-proper as food for the individual insect.

Here we should bear in mind the following consideration: bees of different ages consume different quantities of beebread, nurse-bees and builders taking the most. When, however, food with an excess of protein gets into the honey-crops of foragers and home-bees engaged in storing honey, this probably changes the natural process of feeding for the whole of the colony.

But then, may not the selectiveness possessed by all organisms cause the removal of excessive protein from this "unnatural" food? Perhaps, enveloped with the peritrophic membrane, it is discharged from the alimentary canal, undigested?

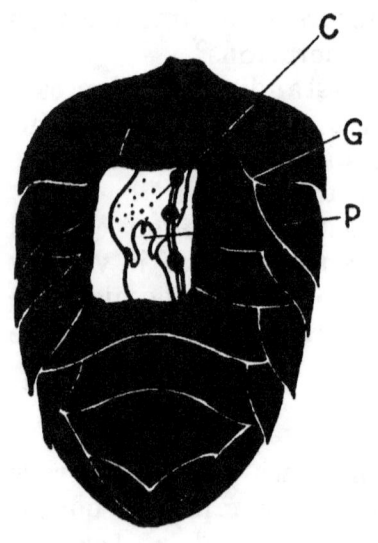

Through this "peep-hole" cut in the chitinous test of the abdomen one can see the honey-crop (C), the stomach-mouth (P) between the honey-crop and the proventriculus, and the ganglia (G). If before the operation a bee is given enough thick syrup with coloured pollen, the observer will see the pollen-grains in the honey-crop and the action of the stomach-mouth detaining them and thus purifying the syrup

By studying at different intervals after feeding the different parts of the alimentary canal—the honey-sack, the intestines, the rectum—and then the haemolymph and the evacuations, one may follow the course of the assimilation of proteins. Analyses have shown that the protein content in the body of a bee fed with the mixture of honey and beebread is one and a half times the normal. This means that man has succeeded in overcoming the first chain of barriers in bee biology and that the stomach-mouth is the breach through which food can be introduced directly and change the development of bee nature.

Perhaps this is exaggerating the importance of the few extra grains of pollen (the absolute magnitudes here are very small) that a bee must digest in order to put into the cells as pure honey the mixture she has taken from the feeder.

Let us recall the modifying effect of habitual excess of highly-nutritious food, or an excess relatively to the wear and tear of the organization from exercise.

The new feeding mixtures consumed and assimilated by bees are examples of that excess of highly-nutritious food which is a powerful cause of variability.

The published results of the tests of feeding with the mixture of honey and beebread state: "While consuming more pollen, the bees spent less honey per unit of brood fed and wax secreted." But this means overcoming another barrier in the defences of the bee-colony: the strictly observed rations for brood-rearing which make for conservatism of bee heredity have been violated.

We may consider this method the first step towards the natural feeding of the brood with artificially changed food.

Professor T. V. Vinogradova worked in the same direction but she fed her bees with yeast fungi and vitamin extracts, instead of the mixture of honey and beebread.

In trial experiments, patches of brood in the comb were fed from one to four drops of additional food (besides that given by the nurse-bees) poured with a curved glass pipette into the cells of brood at fixed intervals.

Young larvae received the additional food in the form of drops on the cell-walls, older larvae ate this food direct from the pipette during the days preceding pupation. Observing this process under the magnifying glass one can distinctly see a larva stretching its mouth to the pipette and sucking the drop from the glass "soother."

The experiments have established that ten drops of yeast received by a larva reduce its passage through the larval phase by one or two days. While the control larvae in the same comb were sealed, as is usual, on the sixth or seventh day, the larvae under experiment were ready for sealing after four and a half days.

The feeding of a queen larva with the yeast fungi and vitamin extracts reduced its development by one day and more. Some of the queen larvae reared on the yeast were ready for sealing by the beginning of the fourth day. The three or four drops of this food reduced the period of development by half!

The most important thing about the experiment was that the queens and workers hatched from the larvae were perfectly sound and vigorous.

But this method of additional feeding is complicated and cannot be used widely. So the task of feeding the brood with artificial protein and vitamin mixtures was simplified, the bees themselves being made to do it.

In early spring twelve colonies were fed sugar syrup to which were added different quantities of ordinary yeast, brewer's yeast grown in a sterilized medium, or of vitamin extract from yeast.

Observation of the development of the colonies, as well as of individual bees, has proved with finality that additional feeding accelerates the development of both larvae and pupae. Students of bee biology use not only mixtures of protein and vitamin foods, but different kinds of flour as protein-substitutes and of malty sugars as carbohydrate substitute.

Not only scientists are engaged in looking for substitute foods for bees. The work done in the Victory Collective Farm, Kanevskaya District, in the Kuban, is becoming increasingly popular among bee-keepers in the Southern regions of the Soviet Union.

On a plot near the collective-farm apiary of about one hundred hives several early and late varieties of water-melons are sown in spring.

A heavy set of fruit results from the pollination by bees and thousands of big water-melons start ripening at the beginning of August, becoming more juicy and sweet with every day. At this season no honey-plants bloom in Kanevskaya District, so when the bee-keeper cuts open a dozen ripe water-melons and leaves them in the sun after crushing their surface, scout-bees discover the sweet bait and soon clouds of foragers suck the thickening juice.

The next day the bees will be found easily disposing of about a quarter of a ton of melons and will bring into the hives 20 to 25 kilogrammes of honey. True, a quarter of a kilogramme of honey per colony per day is not much, but thanks to this slight intake the queens lay more intensively and the honey-stores in the hives increase slowly but surely. Since feeding with water-melon juice continues every day for at least a fortnight (later-ripening varieties come last), each colony collects about three kilogrammes.

Seeing that colonies thus fed build to strength by the autumn and winter well on the water-melon honey, owing to which they have more young bees in the spring and develop better, the collective-farm bee-keepers have water-melon plots near all the eleven apiaries of the collective farm. The experience of the front-rank bee-keepers shows that the conditions of artificial feeding can successfully be made to resemble the natural ones.

If upon the termination of the honey-flow bees are left for some time in big two-body hives, strong colonies continue sending their foragers afield. But the nectaries are empty and the fielders bring home only big loads of pollen. During three or four fine days a colony may fill several combs with beebread from the pollen of late autumnal flowers. Old bees will work themselves to death collecting pollen, but they will not have eaten honey during their last days in vain. When preparing his bees for wintering, the bee-keeper removes the frames with the late beebread, puts them into a big barrel and covers over with honey. The beebread keeps very well if stored in this way. By the end of winter, before the earliest pollen-yielding plants start blooming, the bee-keeper puts one frame of this beebread into each hive.

In colonies given pollen in this way the rearing of brood starts much earlier and the colonies build to strength much more quickly. Besides, some bee-keepers firmly believe that such colonies send no swarms, since bees reared on late pollen are not inclined to swarming.

All these methods may have their shortcomings, but it is easy to see that such practices spell the end of the history of the half-domesticated bee, which man could utilize more or less successfully by adapting himself to her conservative habits, to the immutable laws of the bee community created by nature herself.

While she fed herself, the bee, housed in a hive, could remain independent and the least variable of all domestic animals. But fed with new food prepared according to the bee-keeper's recipe, she will finally become fully domesticated.

Perhaps, now that simple methods of changing bee food for obtaining desirable results have been found, it will be possible to find ways of undermining bee heredity more quickly.

We know that the inhabitants of a hive so fiercely attack the larvae and pupae of strange species that in a few minutes nothing is left of them but dry skins.

Bearing this in mind, we may hope by breeding a queen in a colony which receives protein food only in the shape of young pupae of, for instance, *long-tongued* bumble-bees or *non-stinging* Ceratina, to arm ourselves with a new means of directed breeding.

May we not try to reach this goal another way? While observing the life of bumble-bees in an artificial glass-walled nest, it is interesting to put into it a piece of bee-comb with brood about to emerge. Hatched in a bumble-bee nest, the bees are accepted by the native population and live in peace with them. Of course, the bees behave as is their wont and, bee-like, try to take part in the life of the nest. Bees with injured wings cannot fly, they remain in the nest and soon begin to take nectar for storing from field bumble-bees, to rear the brood, to groom and to feed the queen.

These experiments, interesting from various aspects, cannot but be regarded as one more possibility of bringing about inter-specific hybridization.

Excessive feeding, new conditions of life changing the anatomy and behaviour of individuals and the organization and character of entire colonies, food (vegetative) hybridization preparing new material for further selective breeding, all these will be used to create the first cultivated varieties of bees. The creative activity of man will call to life and develop individual deviations from the prototype, here, too, opening the way to that multitude of different forms which is characteristic of all other domesticated animals and plants. It has taken the progeny of the wild rock-pigeon five thousand years to develop into the Pouters, Carriers, Trumpeters, Frill-Backs, African Owls, Laughers, Swallows, Tumblers, Yacobins, Spots, Nuns, and dozens of other, less known, breeds.

Gooseberry with dozens of its varieties has been cultivated for at least five hundred years. Forest strawberry had been cultivated for 250 years before the different varieties—large-fruit, green, white, red and black—were obtained.

The process of domestication of the honey-bee will not take so long and there will be an infinite number of varieties. If we are not afraid of letting our imagination run away with us, we can fancy in the future non-stinging bees, bees with extra-long proboscides reaching nectar in the deepest flowers, and, particularly, furry bees resembling bumble-bees which, in obedience to training, will pollinate the new plants created by the followers of Michurin, and also specialized wax-producing bees. Bees, probably, will play an important part in the selective breeding of exceptionally viable plant varieties. Perhaps bees will be used even in prospecting: the discovery in the combs of pollen from plants peculiar to areas with certain ores will make it possible to see what is under the ground without crossing the threshold of one's study. For this purpose far-flying varieties without any fixation instinct will be needed, which will visit all flowers that they chance to see.

It is quite possible that bees will help man in solving still more far-reaching and important tasks.

DREAMS AND PLANS

When the Earth Becomes a Garden in Bloom. "Spring"
and "Winter" Bee-Colonies. Perennial Colonies of the
Hymenoptera. The Lesson of the Accrete Roots of Oaks
in Clusters. "Moving Rocks Endowed with Free Energy."
The Force of Advanced Theory.

Let us recall the events from the history of the earth
mentioned at the beginning of this book. Let us remember
that the latest phase in the development of the green robe
of our planet corresponded with the appearance of insects
pollinating flowers. We know how the evolution of plant species was accelerated after the victory of the angiospermae
flower species, and that these species owe their prominent
place in the history of plants to the appearance of pollinating
insects.

Are not these historical facts proof that directed pollination can help botanists, agronomists, selectionists, and
ecologists, as Michurin said, "to do away with time and
call to life beings of the future which otherwise would have
had to wait ages before appearing."

Today, this is being done on a wide scale to agricultural crops, to cultivated plants alone. But the first successful steps have already been made in selective growing
of tree varieties.

What about the future?

Along what lines is life developing? What is dying away
and what is being born? These are questions of the greatest
interest for us.

True, a complete control over the evolution of vegetable life on the earth is a thing of the future, but that future is not so hopelessly far away. Time will provide scientific and technical means for the accomplishment of what
seems today a fantastic task. Then the four-winged pollinators of flowers—the living catalyzers in the process of plant
evolution—will become instruments of natural-scientific
transformation, by means of which the earth will once
again be converted into a garden in bloom, but this time
by man and not by insects.

And this is not all. The zeal and perseverance with which man studies the many-sided organization of the bee-colony are not just the result of his interest in honey and wax, or of his concern about plants being pollinated to ensure a harvest.

Selective animal-breeders dream of bees that will work red clover and alfalfa without special training. Plant-breeders, in their turn, cherish the dream of such varieties of these leguminous plants that will attract bees

The bee-colony reveals to man the real force of vital links unifying a biological species. Starting from the bee-colony, the human mind turns to new and unknown regions of living nature, to their unexplored and inexhaustible resources.

The notion of life, of a living being, is inseparably connected in our minds with an individual growing and developing. This individual may be microscopic or it may be enormous, but it must be born and it must pass its individual course of life, always remaining a distinct being.

In the biological sense of the term, however, life is neither a mass nor a particle of living substance, but a process: a constant process of metabolism with surrounding nature, a process of assimilation and dissimilation.

We can observe and study this process as exemplified not only by individual bees but also by the short-lived, ephemeral spring colonies of solitary bees.

We see its manifestation in the annual colonies of wasps and bumble-bees whose females are revived by the sun after hibernation and germinate like live seeds to new life. Surrounded by new generations, they develop into colonies which produce fruit and spread a new crop of hibernating females, and then die away at the first autumn frosts like annual grains and grasses.

We see its manifestation in the colonies of bees, ants and termites, which may be likened to perennial plants. Such colonies multiply either through living shoots—swarms—as in the honey-bee colony, or through tillering, as in other species, in which new colonies spring up from the mother colony, the root, and live side by side with it. There are colonies bearing fruit every year, from which thousands of males and females fly and crawl away in couples to give rise to new perennial colonies which will grow and unite hundreds of thousands and even millions of individuals they themselves have given birth to.

Considered in this new aspect in those palpitating entities, those "organisms of organisms", life goes on with duality in unity. Life here is at one and the same time the existence of individuals, growing, developing and multiplying, and the existence of their vital organizations, also growing, developing and reproducing and making each individual, perfect in itself, part of a complex and organic structure.

It is very important to grasp what gave rise to those structures, what has reared them and what enables them to develop.

Our quest for an all-embracing answer to these questions takes us from the laboratory, where the phenomena of extracellular metabolism are studied in the colonies of the protozoa, to an old oak grove, where out of thousands of young trees that used to grow in it but a few hundred giant oaks have remained.

If we consider the surviving oaks, we shall see that they are not scattered but grow in more or less clearly defined groups, clusters. These naturally preserved clusters of old oaks also manifest that law of life which has developed out of the biological connections between individuals of the same species.

Acorns germinate much better and the young oaks grow quicker in clusters, which seem so crowded, than if sown one by one in a line, although in the latter case each seedling has a generous allowance of room and food. Thus, life is hard for a lone oak and much easier in a cluster.

But recently the advantages of cluster sowing were seen in the fact that the shade created by plants in a cluster also created a micro-climate beneficial for the development of the seedlings; in the fact that the leaves shed by the trees in autumn covered the ground well, and in a thicker system of roots, better suited for draining the soil.

Experience has proved now that by regulating the number of plants in a cluster, we may exercise directed influence on the nature of their development. If maize, for instance, is grown for green fodder and is to be mowed before flowering it is sown in dense clusters where it grows three or four metres high. Where big cobs are to be obtained, not more than two plants are left in a cluster. In northern districts where period of day-light is exceedingly long it is advisable to establish an artificially-shortened day for young shoots of maize. Plants that have been deprived of light for several hours a day during a month, on being transplanted in clusters in the open air start shooting cobs as early as the third or fourth node, which not only have ample time to develop well but actually ripen and produce seeds suitable for sowing.

Light conditions are of paramount importance for annual plants in the north, but there are other important factors.

Now we know more: the roots of trees in a cluster, oaks for example, become intertwined and accrete, resulting in a communicating, interconnected system of roots.

People see the sturdy trunks of the young oaks, they see the slender branches and lacy crowns, and think that each oak is an independent tree. But underground, the roots of the young trees are intertwined and form an oak "colony" living one common life. If a tree is cut its roots may continue to supply food to its neighbours.

The time is past when the minds of biologists were obsessed with the idea that the law of the jungle reigned supreme in living nature, that general struggle and competition were beneficial for biological species and necessary for their perfection. This idea was born of and supported by the abnormalities of the capitalist mode of life.

In his well-known letter to P. P. Lavrov F. Engels ridiculed those "naturalists, save the mark," who deemed it possible "to reduce the historical development, rich and multiform, to the one-sided and dry formula of 'the struggle for life' which even in nature can be accepted only conditionally." Engels said that "this method condemns itself" and explained that "the interaction of bodies both in living and non-living nature includes harmony and collision, struggle and co-operation."

Michurinist biologists have shown how the harmony and co-operation of a given species with other species can be utilized in the field to bring about conditions for ensuring good crops and also how the struggle and collision between a species and other species can be made use of to protect the crops from pests and disease.

As for the relations between individuals belonging to the same species, these cannot be defined either as a "struggle" or as "mutual assistance," for the life of each individual of a species and of all of them taken together is the life of the species itself.

The new scientific conceptions of the nature of species and the life of species, of the relations between individuals belonging to different species or to the same species have been proved in practice.

As time goes by, science will reveal numerous conditions of the life of species of which we know nothing today as yesterday we knew nothing of the accrete roots of trees underground. The reference is to the conditions drawing domestic animals into herds and droves, and wild animals into packs, the conditions under the influence of which birds form sea-shore colonies and migrating flocks, the conditions bringing about schools of fish in rivers, lakes, seas and oceans. Science will certainly explain why the telenomus gathers in such multitudes under rotten leaves in a forest clearing, why Halictus males congregate on bare branches in the evening and in bad weather, and why a solitary female pigeon cannot lay eggs.

When man has revealed these secrets, big and small, he will be in possession of the key to one of the most wonderful forces of living nature.

An example can be seen in locusts, which are still a scourge beyond the borders of the U.S.S.R.

These Orthoptera migrate almost every year in clouds sometimes covering thousands of square kilometres and weighing millions of tons. Scientists registered one such cloud covering almost 6,000 sq. kilometres, whose weight was calculated to equal that of the copper, lead and zinc extracted by man in a century.

"And this was not one of the biggest clouds," Academician V. I. Vernadsky wrote. He added: "This cloud of locusts expressed in chemical elements and metric tons can be regarded as analogous to a mountain, or rather a moving mountain endowed with free energy. In the face of the variety and greatness of living nature, a cloud of locusts is an unimportant and transient phenomenon. There are phenomena infinitely greater and mightier, such as the coral and calciferous seaweed growths, the live films of plankton stretching unbroken over thousands of square kilometres in the ocean, the floating seaweeds in the Sargasso Sea, the taiga of Western Siberia or the *hylaea* in tropical Africa. Such masses of living matter can be likened to mountains."

To learn to concentrate or to pulverize these moving mountains endowed with free energy (in order to use them or eliminate them) would mean mastering this force, making these masses of living matter obey the will of man. To learn to convert solitary species into species living in biological communities would mean acquiring a new potent means of controlling the nature of insects.

Mutual feeding can be observed in all social insect species: bees, ants and termites, and, to a certain extent, bumble-bees, which live in communities at certain stages of their life-cycle

These prospects, remote, perhaps, but undoubtedly immensely attractive, arise from the experience of controlling the life of the bee-colony as a unit, from the analysis of laws governing bee life. Not until our time did such ideas, which frequently occurred to biologists in the past, assume definite outlines urging naturalists to a deeper study of the problem of species.

For the time being, bee-keepers and students of bee life have simpler tasks to solve. Means must be found of making bees successfully pollinate red clover and alfalfa. Cotton and sunflower fields, too, must become a profitable pasture for bees. This alone will immensely increase annual honey crops in the U.S.S.R. More spirit must be shown in the elaboration of methods for controlling the field activity of bees, making them an obedient tool in the hands of plant-breeders, increasing the force and vigour of varieties. This alone will further increase by millions of centners the yields of various crops cultivated in the fields of the Soviet country.

Bee-keepers have many difficulties ahead of them but progressive biology will help them to overcome all the obstacles in their way.

Theory becomes a concrete force as soon as it becomes the property of the masses, as soon as the masses master it. Every day brings the Soviet people proof that this law formulated by the teachers of working humanity operates in all aspects of the life of Soviet society.

Armed with the teaching of Michurin, millions of tillers of socialist land, workers in the green factory, will continue their searchings, studying and reforming living nature.

With every passing year they will more and more successfully reveal and utilize the forces of nature, enlist the assistance of these forces and direct them in accordance with their own aims.

As I. V. Michurin wrote long ago, this "our common work is a *very important step forward*, it is of world significance, as will be made apparent to all by the future results of this work, the first mighty impulse to which was given by the Revolution that awakened millions of creative minds in the Soviet country."

This creative impulse manifests itself in the everyday work of millions of working people of the Soviet Land, workers in the green factory.

Orderly rows of the future forest belts stretch crisscross over thousands of kilometres; there, in clusters, rise young oaks, spreading their green leaves and growing stronger and stronger as years go by.

A rainbow glitters over the crops in the drops of artificial rain showered by spraying-machines.

The sky high over the steppe is reflected in blue mirrors of reservoirs framed with luxurious greenery. A network of irrigation canals connects vast fields.

Whole expanses are covered by tall maize growing in a dense mass. A tractor proceeds along a field road dragging a number of carts covered with fluffy mounds of maize stalks and leaves. The maize is being taken to the stock-farms, to the silo towers, whose tiled roofs are a prominent feature of the landscape.

Powerful tractors furrow the fields with glittering steel coulters and leave behind wide ribands of soil made fertile by the roots of sown mixed grasses.

Fields of these grasses form an endless carpet and bees sent here by man reach with their proboscides into floret after floret of collective-farm clover.

All this has been done by the hands and minds of the Soviet people, people that were the first in the world to become masters of their own destiny and are the first to become masters of Nature.